A2 UNIT 2

STUDENT GUIDE

CCEA

Physics

Fields, capacitors and particle physics

Ferguson Cosgrove

HODDER
EDUCATION
AN HACHETTE UK COMPANY

Hodder Education, an Hachette UK company, Carmelite House, 50 Victoria Embankment, London EC4 0DZ

Orders

Hachette UK Distribution, Hely Hutchinson Centre, Milton Road, Didcot, Oxfordshire, OX11 7HH

tel: 01235 827827

e-mail: education@hachette.co.uk

Lines are open 9.00 a.m.–5.00 p.m., Monday to Friday. You can also order through the Hodder Education website: www.hoddereducation.co.uk

© Ferguson Cosgrove 2017

ISBN 978-1-4718-6395-0

First printed 2017

Impression number 5 4

Year 2022

This guide has been written specifically to support students preparing for the CCEA A-level Physics examinations. The content has been neither approved nor endorsed by CCEA and remains the sole responsibility of the author.

Cover photo: kasiastock/Fotolia; p.6, PhotosIndia.com LLC/Alamy Stock Photo

Typeset by Integra Software Services Pvt. Ltd, Pondicherry, India

Printed and bound by CPI Group (UK) Ltd, Croydon, CR0 4YY

Contents

Content Guidance

Questions & Answers

■ Getting the most from this book

Sample student answers

Practise the questions, then look at the student answers that follow.

Exam-style questions

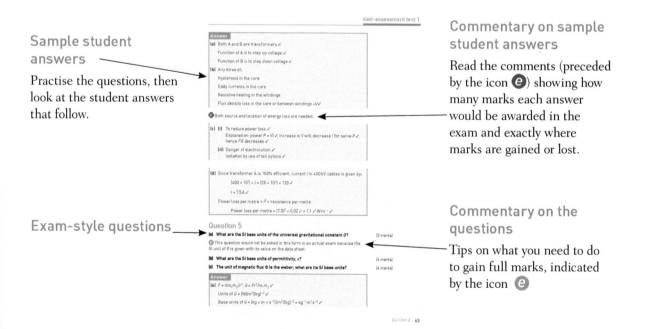

Commentary on sample student answers

Read the comments (preceded by the icon **e**) showing how many marks each answer would be awarded in the exam and exactly where marks are gained or lost.

Commentary on the questions

Tips on what you need to do to gain full marks, indicated by the icon **e**

■ About this book

This book covers the CCEA specification for A2 Physics Unit 2: Fields, capacitors and particle physics. It has two sections:

■ The **Content Guidance** covers A2 Unit 2. It does not have the detail of a textbook but it offers guidance on the main areas of the content and includes worked examples. These examples illustrate the types of question that you are likely to come across in the examination. The exam tips will help you to understand the physics and give you guidance on the core aspects of the subject. They also show how to approach revision and improve your exam technique.

■ The **Questions & Answers** section comprises three self-assessment tests. Answers are provided and there are comments on the specific points for which marks are awarded.

The specification

The CCEA specification is a detailed statement of the physics that is required for the unit assessments and describes the format of the assessments. It can be obtained from the CCEA website at www.rewardinglearning.org.uk.

Symbols, signs and abbreviations

Questions will be set using SI units. You must be familiar with the symbols for quantities relevant to the specification and with the units for these quantities. Questions will assume knowledge of the symbols for the decimal multiples and submultiples shown in the table on p. 56.

Revision tips

The examination is designed to test all the content of the unit. All the questions are compulsory so your revision must address every element.

Break down and learn in detail the content stated in the specification and covered in the Content Guidance of this student guide:

■ experiment descriptions (two named in A2 Unit 2)
■ full definitions
■ equations, in all forms
■ statements of laws and principles

Use past paper questions and those provided in this student guide to ensure that you are familiar with typical questions associated with each topic.

Sequence your notes and corrected example questions in the order of the specification.

Content Guidance

■ Force fields

We are familiar with some of the effects of forces: we exert pushes and pulls on numerous objects. But in everyday situations the force is always 'hands on' or via something rigid along which the pushing force can act or, for a pull, a non-rigid connector such as a rope would also work.

But in physics we also deal with invisible forces which do not make contact with the object they act upon, but can pull or push the object just as readily as if we placed our hands on the object and physically moved it.

We are familiar with examples of the effects of such forces:
- Gravitational — when we jump up off the ground we are pulled back down by this invisible force.
- Electrical — when we hold a rubbed plastic ruler just above small pieces of paper they jump up, lifted by an invisible force.
- Magnetic — when a magnet is held just above small iron pins they jump up, attracted to the magnet by another invisible force.

In an attempt to help visualise these unseen forces we use the concept of a force field. It is a 'map' around the source of the force that shows the pattern of the force in the three-dimensional space around it.

The most familiar and most visual example is that of the field associated with a bar magnet.

If a bar magnet is placed under a sheet of thin card, iron filings sprinkled over the card, then the card tapped gently, the iron filings will move into a definite pattern (Figure 1). A series of lines is observed — the filings have been moved to positions described as magnetic **field lines**, making the magnetic field pattern clearly visible.

A **force field** is the region of space surrounding an object where another object will experience a force due to its related property of mass or charge or its magnetic nature.

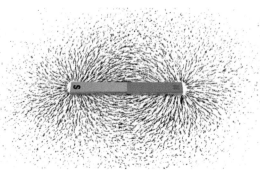

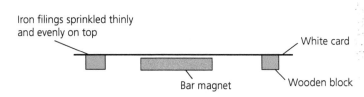

Iron filings sprinkled thinly and evenly on top

White card

Bar magnet

Wooden block

Figure 1 Using iron filings to show the shape of a magnetic field

The forces acting in a field are, like all forces, vectors and so have direction as well as magnitude.

With the iron filings the force lines radiate out from both ends of the bar magnet, but we cannot identify their direction. But if we use a plotting compass to trace out the magnetic force lines (Figure 2), we see that they actually radiate out from one end (N pole) and converge in towards the other (S pole).

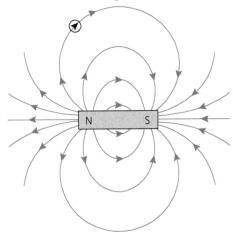

Figure 2 Magnetic field pattern of a bar magnet

The field line arrangement describes the force variation in a field:

- converging field lines — lines that are getting closer together — the field is getting stronger
- diverging field lines — lines that are getting further apart — the field is getting weaker
- parallel field lines — the field strength is constant or uniform

Summary

- A force field is a model to help explain interactions between objects not in direct contact.
- Field lines converge to indicate strengthening, diverge to indicate weakening, and are parallel to indicate constant or uniform field.

■ Gravitational fields

Gravity is something we accept without question, but why is it that when an object is dropped it falls?

Isaac Newton's proposal that an apple falls due to a force of attraction between the apple and the Earth provided the solution to a problem that faced scientists at that time — 'What provides the centripetal force to keep the planets in orbit around the Sun?'

Newton asserted that the force of gravity is the attractive force that acts between all masses, i.e. it is a universal force and not isolated to the situation of objects falling towards the Earth. Newton summarised this in his law of universal gravitation.

The law of universal gravitation states that every particle of matter attracts every other particle with a force that is directly proportional to the product of their masses and inversely proportional to the square of their distance apart.

The law can be expressed by the equation:

$$F = \frac{Gm_1m_2}{r^2}$$

where G is the universal gravitational constant $= 6.67 \times 10^{-11}\,\mathrm{N\,m^2\,kg^{-2}}$.

There is a region surrounding any mass, in which if another mass is placed it will experience a force of **attraction**. We call this the gravitational field.

Worked example

Estimate the gravitational force between two students sitting next to each other in a classroom.

Answer

Assume each student has a mass of 50 kg and there is a distance of 1 m between their centres of mass.

$$F = \frac{Gm_1m_2}{r^2}$$

$$F = \frac{(6.67 \times 10^{-11})(50)(50)}{1^2} \approx 2 \times 10^{-7}\,\mathrm{N}$$

No wonder you are not aware of any attraction!

A gravitational field is a region of space where a particle experiences a force because of its mass.

Knowledge check 1

The radius of the Moon's orbit about the Earth is 3.8×10^8 m. Given that the mass of the Earth is 6.0×10^{24} kg and the mass of the Moon is 7.4×10^{22} kg, calculate the gravitational force of attraction between the Earth and the Moon.

Gravitational field strength

The gravitational force experienced by an object in a particular field will depend on where it is placed and the size of its mass. To allow comparison of one force field with another and one position with another, we use the term gravitational field strength, which is the force acting on unit mass at that point.

We can give a value to the strength of the gravitational field using the equation:

gravitational field strength $E_G = \dfrac{F}{m}$

Gravitational field strength is measured in newtons per kilogram, $\mathrm{N\,kg^{-1}}$.

But F/m is numerically equal to the acceleration of the object, from Newton's second law of motion, $F = ma$. If applied specifically to the Earth's gravitational field and the acceleration of free fall, then:

$$g = E_G = \frac{F}{m} = \frac{GM_E}{r^2}$$

Therefore gravitational field strength may also be measured in $\mathrm{m\,s^{-2}}$.

A gravitational field can be represented by sketching field lines around a mass. The field lines show the direction of the force that acts on the mass.

Gravitational field strength at a point is defined as the force per kilogram acting on a mass placed at that point in the field.

Knowledge check 2

Find the gravitational field strength at the surface of the Sun, taking the radius to be 7.0×10^8 m and the mass to be 2.0×10^{30} kg.

If we consider the Earth as a uniform sphere then its field pattern, shown in Figure 3, is a set of converging, symmetrically spaced, radial lines.

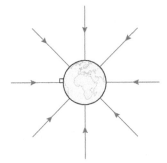

Figure 3 Gravitational field pattern of the Earth

Note the lines are perpendicular to the Earth's surface. Therefore close to the Earth's surface, where we view the surface as horizontal, the field lines are approximately parallel and so the field can be viewed as uniform, see Figure 4.

Earth's surface

Figure 4 Gravitational field lines close to the Earth's surface

> **Exam tip**
>
> Note that the gravitational field strength E_G is a vector quantity whose direction is that of the force F.

Mass of the Earth

Applying Newton's law of universal gravitation to an object on the Earth's surface:

$$E_G = g = 9.81 \text{ m s}^{-2} = F/m$$

Taking the Earth to be a uniform sphere allows us to treat the Earth as a point mass with a separation from an object on its surface equal to its radius:

$$E_G = g = \frac{GM_E}{r_E^2}$$

And so:

$$M_E = \frac{gr_E^2}{G}$$

Taking the radius of the Earth as 6400 km,

$$M_E = \frac{9.81 \times (6.4 \times 10^6)^2}{6.67 \times 10^{-11}}$$

$$M_E = 6.0 \times 10^{24} \text{ kg}$$

Note that this calculation of mass is not restricted to the Earth, nor to an object on the surface. We just need to know the field strength ('g' value) at that point and the separation of that point from the centre of mass of the attracting body.

> **Exam tip**
>
> Gravitational forces are in general very small and insignificant. For example two 1 kg masses placed 1 m apart experience a gravitational attraction equal to 6.67×10^{-11} N (the value of G). It is only when dealing with large masses, such as planets, that the force is considerable and plays a major role.

Planetary motion

In the early 1600s Johannes Kepler, using records of astronomical observations made by his mentor Tycho Brahe, published three laws on planetary motion:

- Law 1: the planets describe ellipses about the Sun as one focus.
- Law 2: the line joining the Sun and the planet sweeps out equal areas in equal times.
- Law 3: the square of the period of revolution of the planet is proportional to the cube of its mean distance from the Sun.

Using Newton's law of universal gravitation and approximating a planet to move in a circle around the Sun, we can see that Kepler's third law is consistent with the theory.

Assuming it is the gravitational force that provides the centripetal force, then:

$$F_G = F_C$$

$$F_G = \frac{GM_SM_P}{r^2}$$

and

$$F_C = M_Pr\omega^2 \text{ (where } \omega = 2\pi/T)$$

Therefore:

$$\frac{GM_SM_P}{r^2} = M_Pr\omega^2 = M_Pr\left(\frac{2\pi}{T}\right)^2$$

and

$$T^2 = \left(\frac{4\pi^2}{GM_S}\right)r^3$$

where π, G and M_S are constant.

So:

$$T^2 \propto r^3$$

Kepler's third law is thus derived using Newton's law of gravitation.

This theory can also be used to determine the mass of the attracting body, in this case the Sun.

$$M_S = \frac{r^3 4\pi^2}{GT^2}$$

where $r = 1.5 \times 10^{11}$ m, the distance between the Sun and the Earth, and $T \approx 365$ days, the period of one revolution.

$$M_S = \frac{(1.5 \times 10^{11})^3 \times 4 \times \pi^2}{(6.67 \times 10^{-11}) \times (365 \times 24 \times 60 \times 60)^2}$$

$$M_S = 2.0 \times 10^{30} \text{ kg}$$

Geostationary satellites

Satellites are kept in orbit in the Earth's gravitational field by the gravitational attraction force of the Earth.

It is particularly useful to have a satellite that will stay over the same place on the Earth while the Earth rotates, and this is referred to as a **geostationary** or 'parking' orbit. The satellites used to relay television programmes, for example, have to be in fixed positions relative to the Earth, so that domestic satellite dishes don't have to 'track' the satellite across the sky.

There are a unique set of conditions that must be fulfilled if such an orbit is to be achieved:

- The satellite must circle the Earth in the plane of the equator.
- The direction of rotation must be the same as that of the Earth.
- The period of rotation must be the same as that of the Earth as it turns about its axis, i.e. 24 hours.

Worked example

a What is the height of a geostationary satellite's orbit above the Earth's surface? (Radius of Earth = 6.4×10^6 m)

b What is the speed of the geostationary satellite?

Answer

a Let R be the distance of the satellite, mass m_s, from the centre of the Earth, mass m_E, and v its speed in orbit. $F_G = F_C$, so:

$$\frac{G m_E m_S}{R^2} = \frac{m_S v^2}{R}$$

Using $G m_E = g r_E^2$ where r_E is the radius of the Earth:

$$\frac{g r_E^2 m_S}{R^2} = \frac{m_S v^2}{R}$$

Therefore:

$$v^2 = \frac{g r_E^2}{R}$$

If T is the period of the satellite in its orbit, then $v = 2\pi R/T$. Therefore:

$$\frac{4\pi^2 R^2}{T^2} = \frac{g r_E^2}{R}$$

and

$$R^3 = \frac{T^2 g r_E^2}{4\pi^2}$$

With $T = 24$ hours, $r_E = 6.4 \times 10^6$ m, $g = 9.81$ m s^{-2}:

$$R = [(24 \times 60 \times 60)^2 \times 9.81 \times (6.4 \times 10^6)^2/4\pi^2]^{\frac{1}{3}} = 4.24 \times 10^7 \, \text{m}$$

The height above the Earth's surface of the geostationary satellite is
$R - r_E = 3.6 \times 10^7$ m.

b In a circular orbit, the speed of the satellite $v = 2\pi R/T$.

$$v = \frac{(2\pi \times 4.24 \times 10^7)}{(24 \times 60 \times 60)} = 3.1 \times 10^3 \, \text{m s}^{-1}$$

Summary

- Newton's law of universal gravitation states that every particle of matter attracts every other particle with a force which is directly proportional to the product of their masses and inversely proportional to the square of their distance apart: $F = Gm_1m_2/r^2$.
- A gravitational field is the region of space, surrounding a mass, in which if another mass is placed it will experience a force of attraction.

- Gravitational field strength at a point is the force per unit mass acting on a mass placed at that point in the field: $E_G = F/m = g$.
- Kepler's third law states that the square of the period of revolution of a planet is proportional to the cube of its mean distance from the Sun.
- Geostationary satellites have the same period of rotation as the Earth, move in the same direction as the Earth and are above the equator.

■ Electric fields

Just as every mass creates a gravitational field around itself, every charged particle creates an **electric field** around itself.

Once again we can visualise an electric field by sketching field lines around a charge (Figure 5). Field lines always show the direction of the force that would be exerted on a **positive charge** placed at that position.

An **electric field** is a region of space where a particle experiences a force because of its charge.

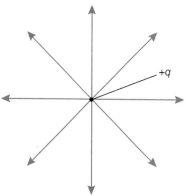

+q

Figure 5 Radial electric field around a point positive charge

Around a point charge the field is radial, with the lines resembling the spokes of a bicycle wheel. We can assess the relative field strength at different places in the field: near the point charge the field lines are more concentrated and so the field is strong; further away from the point charge the lines are spreading out, indicating a weakening field or diminishing field strength.

Coulomb's law

Just as Newton was able to explain the existence of the gravitational forces between masses, an equivalent explanation and law for the case of charged particles was needed.

In 1785, Charles Coulomb, a French scientist, was the first to investigate the magnitude of the force between charged spheres and to formulate a law, called Coulomb's law.

The law may be stated in mathematical terms as:

$$F \propto \frac{q_1 q_2}{r^2}$$

where F is the electric force, q_1 and q_2 the point charges and r is their distance apart.

Expressing the law as an equation requires the introduction of a constant of proportionality, k:

$$F = \frac{k q_1 q_2}{r^2} \text{ where } k = \frac{1}{4\pi\varepsilon}$$

The 4π appears because of SI units, to ensure numerical equality. The constant ε is a property of the medium between the point charges, called the **permittivity** of the medium. In general:

$$F = \frac{q_1 q_2}{4\pi\varepsilon r^2}$$

If the medium is a vacuum, then the permittivity constant is written as ε_0, which can be experimentally shown to have the value $8.85 \times 10^{-12} \, \text{F m}^{-1}$, and so $k = 1/(4\pi\varepsilon_0) = 8.99 \times 10^9 \, \text{F}^{-1} \text{m}$.

Changing the medium will change the value of the permittivity and consequently the force produced. The permittivity of air is about 1.005 times that of a vacuum. For most purposes we approximate the value of ε_0 for that of air. The permittivity of water is about 80 times that of a vacuum. So the force between charges situated in water is 80 times less than if they were situated the same distance apart in a vacuum. (It is for this reason that common salt dissolves in water. The electrostatic forces of attraction between the positive sodium ions and the negative chloride ions, which keep the solid crystal structure in equilibrium, are reduced considerably by the water and the solid structure collapses.)

Coulomb's law states that the force between two point charges is directly proportional to the product of the charges and inversely proportional to the square of their separation.

Exam tip

It is impossible not to see the parallel to Newton's law of universal gravitation, but note in the electrical case the force is not always attractive; it can be repulsive, as in the worked example of two like charges. The force between two opposite charges, positive and negative, will be attractive. An attractive force will therefore have a negative sign.

Calculate the force between two small spheres carrying charges of $+2 \times 10^{-8}\,\text{C}$ and $+4 \times 10^{-8}\,\text{C}$ in a vacuum, the distance between their centres being 5.0 cm.

Answer

Using $k = 8.99 \times 10^9 \approx 9 \times 10^9\ \text{F}^{-1}\text{m}$,

$$F = \frac{kq_1q_2}{r^2} = \frac{(9 \times 10^9) \times (2 \times 10^{-8}) \times (4 \times 10^{-8})}{(5 \times 10^{-2})^2}$$

$$F = 2.9 \times 10^{-3}\,\text{N}$$

Note that as both charges are positive, this will be a force of **repulsion**.

Calculate the force between two electrons $1.2 \times 10^{-10}\,\text{m}$ apart in a vacuum.

The **electric field strength** at a point is defined as the force per coulomb acting on a charge at that point in the field.

Electric field strength due to a point charge

The strength of the electric field at a point, or electric field strength E_E, is the force that would be exerted on a unit charge placed at that point.

$$E_E = \frac{F}{q}$$

Since the force between two point charges (in a vacuum), q_1 and q_2, is $F = q_1q_2/4\pi\varepsilon_0 r^2$, the electric field strength in the radial field of a point charge q is:

$$E_E = \frac{q}{4\pi\varepsilon_0 r^2} = \frac{kq}{r^2}$$

Electric field strength is measured in newtons per coulomb, N C^{-1}.

The direction of the vector E_E is the direction of the force on a **positive** charge placed in the field.

Uniform electric field

It is possible to create a uniform electric field between oppositely charged parallel conductors, such as the plates of a capacitor. In between the plates the field strength is independent of position and just depends on the separation of the plates, d, and the potential difference, V, between them.

To show that the field strength is constant, the field lines are drawn perpendicular to the plates, parallel and equally spaced (Figure 6).

Consider a charge q moved from one plate to the other. The work done on the charged particle, W, is given by:

$$W = F \times d$$

From the definition of potential difference we know that $W = q \times V$.

Be careful not to get confused with your 'q's! Check you know whether q is the charge creating the field or the charge that is acted on by a field.

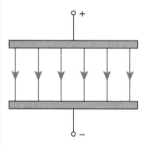

Figure 6 A uniform electric field

Between the parallel conductors the field strength is the same at all points, therefore the force on a charged particle is the same no matter where it is placed in the field.

Therefore:

$$F \times d = q \times V$$

Rearranging:

$$\frac{F}{q} = \frac{V}{d}$$

And so, as $E_E = F/q$,

$$E_E = \frac{V}{d} \text{ for a \textbf{uniform} field}$$

Exam tip

Note the equivalence of the units $N\,C^{-1}$ and $V\,m^{-1}$.

Worked example

An electron is placed between two metal plates, which are 12 cm apart in a vacuum. A potential difference of 150 V is applied across the plates. Calculate the force on the electron.

Answer

$$E_E = \frac{V}{d} = \frac{150}{0.12} = 1250\,V\,m^{-1}$$

$$q_e = e = -1.6 \times 10^{-19}\,C$$

$$F = qE_E = (-1.6 \times 10^{-19}) \times 1250 = -2.0 \times 10^{-16}\,N$$

The force will have a magnitude of $2.0 \times 10^{-16}\,N$, and the electron will be attracted towards the positive plate.

Similarities and differences between gravitational and electric fields

Gravitational and electric fields are both governed by laws that are 'inverse square' in type. The differences arise because there are two types of charge but only one type of mass. Consequently the gravitational force is always attractive, whereas the electric force may be attractive or repulsive (Table 1). Also, the gravitational force is unaffected by the medium between the masses, whereas the force between charges is dependent on the separating medium between charges.

Knowledge check 4

Show that the units of $N\,C^{-1}$ and $V\,m^{-1}$ are equivalent.

Table 1 Comparing the gravitational force and the electric force

	Gravitational force	Electric force
Type of force	Always attractive	Attractive or repulsive
Force law	$F = Gm_1m_2/r^2$ Newton's law of gravitation	$F = q_1q_2/4\pi\varepsilon r^2$ Coulomb's law
Proportional to	Product of masses	Product of charges
Proportional to	$1/r^2$	$1/r^2$
Constant of proportionality	G A universal constant, so independent of the medium	$1/4\pi\varepsilon$ A constant that depends on the permittivity of the intervening medium
(Radial) field strength around a point or uniform sphere	$E_G = g = Gm/r^2$	$E_E = q/4\pi\varepsilon r^2$

A hydrogen atom has a radius of about 5.3×10^{-11} m. It consists of a single electron and a single proton. Calculate and compare the size of the gravitational force and the electrostatic force between the particles.

Answer

$m_p = 1.67 \times 10^{-27}$ kg, $m_e = 9.11 \times 10^{-31}$ kg, $e = 1.6 \times 10^{-19}$ C, $1/4\pi\varepsilon_0 = 9.0 \times 10^9$ F^{-1} m, $G = 6.67 \times 10^{-11}$ N m^2 kg^{-2}

$$F_E = \frac{q_1 q_2}{4\pi\varepsilon_0 r^2} = \frac{(-1.6 \times 10^{-19})(1.6 \times 10^{-19})(9.0 \times 10^9)}{(5.3 \times 10^{-11})^2} = -8.2 \times 10^{-8} \text{ N}$$

$$F_G = \frac{G m_1 m_2}{r^2} = \frac{(6.67 \times 10^{-11})(1.67 \times 10^{-27})(9.11 \times 10^{-31})}{(5.3 \times 10^{-11})^2} = 3.6 \times 10^{-47} \text{ N}$$

The electrostatic force is about 2×10^{39} times larger than the gravitational force. (The gravitational force is always very weak, unless one of the masses is very large, such as a planet.)

Exam tip

The 'sign' of the force evaluated can be misleading, e.g. between opposite charges we will obtain a −ve force and we know the force to be attractive, but between masses we calculate a +ve force, yet we know it to be attractive! It is best to ignore the sign and apply the rules of physics to the situation. (An alternative is to add a negative sign in Newton's law of gravitation.)

Summary

- An electric field is a region of space where a particle experiences a force because of its charge.
- The field lines show the direction of the force that would be exerted on a positive charge placed at that position.
- The force between two point charges is proportional to the product of the charges and inversely proportional to the square of the distance between them. This is called Coulomb's law: $F = kq_1 q_2/r^2$, where $k = 1/4\pi\varepsilon_0$.

- The permittivity of free space, $\varepsilon_0 = 8.85 \times 10^{-12}$ F m^{-1}, and $k = 1/4\pi\varepsilon_0 = 8.99 \times 10^9$ F^{-1} m.
- Electric field strength is the force per coulomb that would be exerted on a charge placed at that point: $E = F/q$.
- The electric field strength at a point in the field of an isolated point charge is given by $E = kq/r^2$.
- The electric field between parallel plates is uniform. The field strength is given by $E = V/d$.

Capacitors

Capacitors are important components in electric circuits; they are used in tuning circuits, smoothing circuits and timing circuits.

A capacitor is a device that can store charge and, as a consequence, energy. Effectively, all capacitors consist of a pair of conducting plates separated by an insulator. The insulator is called a dielectric; typically it may be polystyrene, paper, oil or air.

The amount of energy stored by a capacitor depends on how much charge is moved onto the plates. The plates need to have quite a large area with a small separation. In one type of capacitor this is achieved by taking two long strips of metal foil, separated by a thin insulator and then rolled up like a Swiss roll (Figure 7).

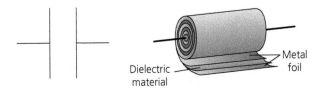

Figure 7 Capacitor symbol and construction

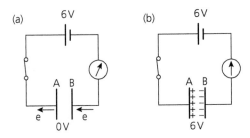

Figure 8 Charging a capacitor

Figure 8 shows a parallel-plate capacitor connected to a battery. When the battery is first connected there is a *momentary* flow of current. This is the result of electrons being deposited on plate B by the action of the negative terminal, while at the same time electrons are being drawn from plate A by the positive terminal of the battery (Figure 8a). As the charge builds up on the plates, one negative, the other positive, the potential difference across the plates rises. Charge will continue to flow until the potential difference across the capacitor is equal to that across the battery (Figure 8b).

Note that during the charging process no charge has crossed the insulating gap.

Capacitance

The measure of the extent to which a capacitor can store charge is called its capacitance. This is defined by the equation:

$$C = \frac{Q}{V}$$

where C = the capacitance, Q = the magnitude of the charge on either plate and V = the potential difference between the plates.

The farad is the SI unit of capacitance, symbol F. However, one farad is too large a value for practical situations. The approximate range of capacitance used experimentally is from the picofarad, pF (10^{-12} F), to the microfarad, µF (10^{-6} F).

Energy stored in a capacitor

When charging a capacitor, the addition of electrons to the negative plate involves doing work against the repulsive forces of the electrons that are already there. Similarly, the removal of electrons from the positive plate requires work to be done against the attractive forces of the positive charges on the plate. The work that is done is stored as electrical potential energy in the capacitor.

Exam tip

The capacitor is said to store a charge Q, although if we add up the charges on both plates and take account of their sign, $(+Q) + (-Q) = 0$, the net charge is zero. Also, as many electrons move off one plate as move onto the other. But, more importantly, the charged capacitor has an increased potential energy.

Note that while the capacitor has charge $+Q$ on one plate and $-Q$ on the other, the charge stored is Q **not** $2Q$.

Capacitance is the charge stored per unit potential difference between the capacitor plates.

Exam tip

Don't mix up your 'C's. The letter C is used as an abbreviation for the unit of charge, the coulomb. The italic capital C is used as the symbol for capacitance.

The farad, F, is defined as one coulomb per volt.

From $C = Q/V$, we get $V = Q/C$, so a graph of V against Q for a capacitor will be a straight line through the origin with gradient $1/C$ (Figure 9).

Consider a small charge, Δq, added to the capacitor when it has a potential difference of V. Then the work done will be $\Delta W = V'\Delta q$, which is the area of the thin strip shown in Figure 9. The total area beneath the line up to charge Q can be thought of as the sum of the areas of many such thin strips, and this total area represents the total work done in charging the capacitor up to that charge Q. This total work done equals the energy stored by the capacitor.

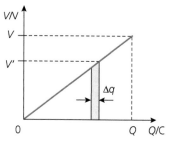

Figure 9 Determining the energy stored in a charged capacitor

Work done in adding $\Delta q = V'\Delta q$ = area of strip

Energy stored by capacitor E_C = total area of all strips

= area of triangle = $\frac{1}{2}QV$

Since $C = Q/V$, this expression can be written in different forms:

$$E_C = \tfrac{1}{2}QV = \tfrac{1}{2}CV^2 = \frac{Q^2}{2C}$$

Capacitors in parallel and series

In circuits, just as resistors can be connected in series and parallel, the same applies to capacitors. We want to find the value of the effective capacitance of capacitors when arranged in series and in parallel.

Parallel capacitors

In Figure 10, the three capacitors of capacitance C_1, C_2 and C_3 are connected in parallel.

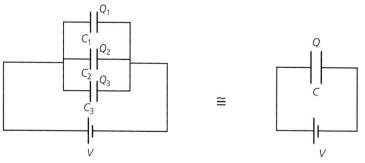

Figure 10 Capacitors in parallel

Key points you need to know about the circuit:
- There is the same potential difference V across each capacitor when in parallel.
- The total charge stored on the three capacitors $Q = Q_1 + Q_2 + Q_3$.

Knowledge check 5

The charge on a certain capacitor is $4.0 \times 10^{-3}\,C$ when the potential difference across it is $50\,V$. Calculate the capacitance of this capacitor.

Knowledge check 6

Calculate the energy stored by a $470\,\mu F$ capacitor at a potential of $12\,V$.

Using $Q = CV$ for each capacitor:

$Q_1 = C_1V, \quad Q_2 = C_2V, \quad Q_3 = C_3V$

Therefore:

$Q = C_1V + C_2V + C_3V = (C_1 + C_2 + C_3)V$

So a single capacitor storing charge Q when the potential difference across its plates is V must have a capacitance:

$C = \dfrac{Q}{V} = C_1 + C_2 + C_3$

The **total capacitance** C for capacitors placed **in parallel** is equal to **the sum of the individual capacitances**.

A 250 μF capacitor in parallel with a 500 μF capacitor is connected to a 12 V supply. Calculate:

a   the total capacitance of the combination
b   the potential difference across each capacitor
c   the charge stored on each capacitor

Answer

a   When in parallel $C = C_1 + C_2 = 250 + 500 = 750\,\mu F$
b   When in parallel the potential difference across each capacitor is the same, and equal to the p.d. of the supply, 12 V.
c   Using $Q = CV$:
   For the 250 μF capacitor:

   $Q = (250 \times 10^{-6}) \times 12 = 3.0 \times 10^{-3}\,C$

   For the 500 μF capacitor:

   $Q = (500 \times 10^{-6}) \times 12 = 6.0 \times 10^{-3}\,C$

Series capacitors

In Figure 11, the three capacitors of capacitance C_1, C_2 and C_3 are connected in series.

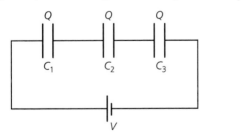 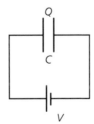

Figure 11 Capacitors in series

Key points you need to know about the circuit:

- The potential difference across the combination is the sum of the potential differences across the individual capacitors: $V = V_1 + V_2 + V_3$.
- A charge $+Q$ on capacitor C_1 will induce a charge $-Q$ on its other plate. This negative charge has been drawn from C_2 leaving it with charge $+Q$. This in turn induces a charge $-Q$ on the opposite plate and so on. The net result is that **each capacitor stores a charge Q and the total charge stored is also Q.**

Using $Q = CV$ in the form $V = Q/C$ for each capacitor:

$$V_1 = \frac{Q}{C_1}, \quad V_2 = \frac{Q}{C_2}, \quad V_3 = \frac{Q}{C_3}$$

Therefore:

$$V = \frac{Q}{C_1} + \frac{Q}{C_2} + \frac{Q}{C_3} = Q\left(\frac{1}{C_1} + \frac{1}{C_2} + \frac{1}{C_3}\right)$$

So, for a single capacitor storing charge Q when the potential difference across its plates is V:

$$\frac{V}{Q} = \frac{1}{C} = \frac{1}{C_1} + \frac{1}{C_2} + \frac{1}{C_3}$$

Capacitors connected **in series** have a **total capacitance** C given by:

$$\frac{1}{C} = \frac{1}{C_1} + \frac{1}{C_2} + \dots$$

Exam tip

Note that the equation for capacitors in series is similar to that for resistors in parallel, and the equation for capacitors in parallel is similar to that for resistors in series.

Worked example

A 250 µF capacitor in series with a 500 µF capacitor is connected to a 12 V supply. Calculate:

a the total capacitance of the combination
b the charge stored on each capacitor
c the potential difference across each capacitor

Answer

a When in series:

$$\frac{1}{C} = \frac{1}{C_1} + \frac{1}{C_2} = \frac{1}{250} + \frac{1}{500} = \frac{3}{500}$$

This gives $C = 166\,\mu\text{F}$.

b The charge stored on each capacitor is equal to the charge stored by the single equivalent total capacitance.

$$Q = CV = (166 \times 10^{-6}) \times 12 = 2.0 \times 10^{-3}\,\text{C}$$

Exam tip

Note that the total capacitance of capacitors in series is *smaller* than the smallest individual capacitance.

c Using $V = Q/C$:

For the 250 μF capacitor:

$$V = \frac{(2.0 \times 10^{-3})}{(250 \times 10^{-6})} = 8.0\,V$$

For the 500 μF capacitor:

$$V = \frac{(2.0 \times 10^{-3})}{(500 \times 10^{-6})} = 4.0\,V$$

Note that the p.ds add up to 12V, as expected.

Knowledge check 7

A 200 μF capacitor and a 300 μF capacitor are connected, first in series to a 6V supply and then in parallel with the same supply. Which arrangement will store more charge?

Connecting two capacitors

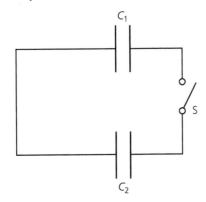

Figure 12 Connecting two capacitors

Figure 12 shows two charged capacitors of capacitance C_1 and C_2. When the switch S is closed, charge will flow from one capacitor to the other, until the potential difference across each capacitor is the same. In order to calculate the final values of charge, potential difference and energy of each capacitor we must note that:

- the total amount of charge does not change
- the final potential difference across each capacitor is the same
- the capacitors are effectively in parallel, so total $C = C_1 + C_2$

Unless the potential difference across the capacitors is equal to begin with, the total energy stored by the capacitors, when they are joined, will decrease. Energy is lost as heat in the connecting leads when charge flows from one capacitor to the other.

Exam tip

When two isolated capacitors are connected as shown in Figure 12, the fact that the final p.d. across each is equal confirms that this is a parallel combination.

Worked example

A 4 μF capacitor is charged by a 60V supply and is then connected across an uncharged 20 μF capacitor. Calculate:

a the final potential difference across each capacitor
b the final charge on each capacitor
c the initial energy and final energy stored by the capacitors

Answer

a The initial charge on the $4\,\mu F$ capacitor is:

$$Q = CV = (4 \times 10^{-6}) \times 60 = 2.4 \times 10^{-4}\,C$$

This is also the final total charge, as none is lost from the system.

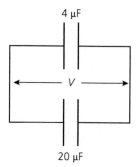

Figure 13

When connected as shown in Figure 13, the capacitors will have a common final p.d. V, therefore they are effectively in parallel and so $C_{tot} = C_1 + C_2$.

$$C_{tot} = 4 + 20 = 24\,\mu F$$

Applying $V = Q/C_{tot}$ to the combination:

$$V = \frac{(2.4 \times 10^{-4})}{(24 \times 10^{-6})} = 10\,V$$

b For the $4\,\mu F$ capacitor:

$$Q = CV = (4 \times 10^{-6}) \times 10 = 4 \times 10^{-5}\,C$$

For the $20\,\mu F$ capacitor:

$$Q = CV = (20 \times 10^{-6}) \times 10 = 2 \times 10^{-4}\,C$$

c The energy of a charged capacitor is given by $E = \frac{1}{2}CV^2$.

initial energy $= 0.5 \times (4 \times 10^{-6}) \times 60^2 = 7.2 \times 10^{-3}\,J$

final energy $= [0.5 \times (4 \times 10^{-6}) \times 10^2] + [0.5 \times (20 \times 10^{-6}) \times 10^2]$
$= 1.2 \times 10^{-3}\,J$

Charge and discharge of capacitors

Figure 14 shows a circuit that can be used to investigate the nature of the charging or discharging of a capacitor. The inclusion of a resistor in series with the capacitor will reduce the current and slow the process, of both charge and discharge, to allow observation or measurement.

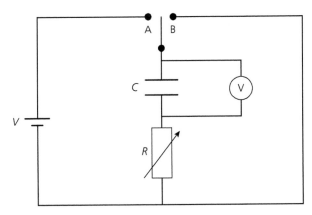

Figure 14 Experimental arrangement for charging and discharging a capacitor through a resistor

When the two-way switch is moved to position A the capacitor will charge up, until the potential difference between its plates is equal to the e.m.f. of the supply, and when the switch is moved to B the capacitor will discharge. The potential difference across the capacitor can be recorded, from the voltmeter, at regular intervals of time as the capacitor charges and discharges. The value of resistor R can be altered so that the time for the process allows accurate observation.

It is also possible to place an ammeter in series with the capacitor to show the variation in the charge and discharge current; note there will be a change in current direction.

Figure 15 shows the variation with time of the potential difference V and current I during the charging and discharging of a capacitor through a resistor.

> **Exam tip**
>
> During discharge of a capacitor the excess electrons on the negative plate are able to move onto the positive plate where there is a deficiency of electrons.

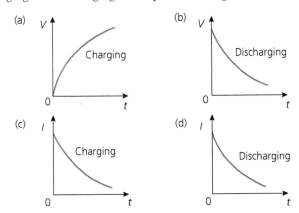

Figure 15 Exponential pattern of potential difference and current during charging (left) and discharging (right) of a capacitor

The potential difference rises and falls exponentially during charging and discharging respectively. The current falls exponentially during both charging and discharging.

The exponential variation is a consequence of the changing ease with which electrons can move onto and away from the capacitor plates. For example, consider charging — initially there are no excess electrons on the plate of an uncharged capacitor, so the rate of flow of electrons (current) will be large. As electrons are deposited on the plate they will repel further additional electrons, so the rate of flow of charge (current) progressively decreases. More detailed practical investigation confirms that the change is exponential in type.

The mathematical representation of the **discharge** of a capacitor of capacitance C through a resistor of resistance R is:

$$Q = Q_0 e^{-t/CR}$$

Q_0 is the initial charge at time $t = 0$ and Q is the charge at time t.

Since $Q = CV$:

$$CV = CV_0 e^{-t/CR}$$

Cancelling C:

$$V = V_0 e^{-t/CR}$$

This is the equation of graph (b) in Figure 15.

Since $V = IR$,

$$IR = I_0 R e^{-t/CR}$$

Cancelling R:

$$I = I_0 e^{-t/CR}$$

This is the equation of graph (d) in Figure 15.

Time constant

As discharge of a capacitor takes place, the exponential curves (b) and (d) in Figure 15 get closer to the time axis, but never actually touch it. So we cannot state a time for total discharge. Within the exponential term $e^{-t/CR}$, the rate at which the capacitor discharges through a resistor depends on the values of C and R. When the product CR is large the decay is slow and vice versa. It is this product that we use to compare the performance of different circuits.

CR is known as the time constant of the circuit. It has the unit of time, and is measured in seconds. Its symbol is τ.

To find the potential difference across the capacitor after a time $\tau = CR$, substitute $t = \tau = CR$ into the equation $V = V_0 e^{-t/CR}$:

$$V = V_0 e^{-CR/CR} = V_0 e^{-1} = 0.37V_0$$

Thus over a time interval equal to one time constant, the charge, the voltage and the current will fall to 37% of their initial value.

> **Exam tip**
>
> The equation for the increase in charge and hence potential difference (graph (a) in Figure 15) when **charging** a capacitor is $Q = Q_0(1 - e^{-t/CR})$. You will not be required to use this equation.

> The time constant of a capacitor circuit is the time for each of the variables Q, V and I to decrease to $1/e$ or 37% of its initial value during discharge.

> **Knowledge check 8**
>
> Show that the time constant, $\tau = CR$, has the unit second.

Worked example

A $2.0\,\mu F$ uncharged capacitor is connected in series with a $5\,M\Omega$ resistor, a $12\,V$ battery and a switch.

a What is the initial current when the switch is closed?

b What is the time constant for this circuit?

c What is the current $3.0\,s$ later?

Answer

a $I_0 = \dfrac{V}{R} = \dfrac{12}{(5 \times 10^6)} = 2.4 \times 10^{-6}\,A$

b $\tau = CR = (2 \times 10^{-6}) \times (5 \times 10^6) = 10\,s$

c $I = I_0 e^{-t/CR}$ applies to both charging and discharging of the capacitor, see graphs (c) and (d) in Figure 15.

$I = 2.4 \times 10^{-6} \times e^{-(3.0/10)} = 1.8 \times 10^{-6}\,A$

Measuring the time constant

Using the circuit shown in Figure 14 and the procedure described, values of the potential difference during discharge and the corresponding times can be directly plotted and the exponential curve drawn. Then, τ, the time corresponding to $0.37V_0$, can be read using the curve, see graph (b) of Figure 15. It is good experimental practice to obtain more than one value for the time constant and calculate an average; for example, 2τ, the time corresponding to $(0.37 \times 0.37)V_0 = 0.137V_0$, could also be read off the graph.

An alternative is to plot a graph of $\ln V$ on the y-axis against time t on the x-axis, as in the following worked example.

$V = V_0 e^{-t/CR}$

So

$\ln V = \ln V_0 + (-t/CR)\ln e$

$\ln V = -(1/CR)t + \ln V_0$

Comparing with

$y = mx + c$

we see that this should be a straight line and the magnitude of the gradient will equal $1/CR = 1/\tau$, the inverse of the time constant.

Worked example

A capacitor is charged to a voltage of 10 V and then discharged through a 100 kΩ resistor. Measurements of the potential difference across the capacitor, taken every 10 seconds, are shown in Table 2.

Table 2

t/s	0	10	20	30	40	50	60
p.d./V	10.0	6.3	4.0	2.6	1.6	1.0	0.7
ln(p.d./V)	2.30	1.84	1.39	0.96	0.47	0.00	−0.36

a Plot a graph of potential difference against time for the discharging capacitor.
b Plot a graph of ln V against time for the discharging capacitor.
c Use your graphs to determine values for the time constant for the circuit.
d Hence calculate a value of the capacitance of the capacitor.

Answer

a

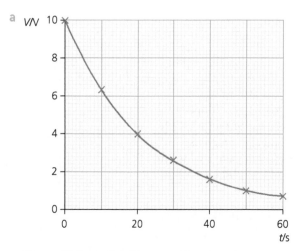

Figure 16 Potential difference V against time

b

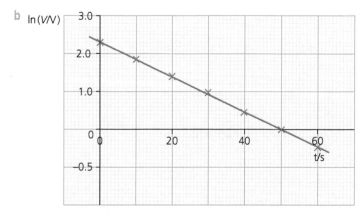

Figure 17 ln V against time

c $V_0 = 10\,V$, $0.37V_0 = 3.7\,V$

When $V = 3.7\,V$ the time elapsed is τ, which, from Figure 16, is about 22 s.

Gradient of graph in Figure 17 = $-2.3/50 = -4.6 \times 10^{-2}\,s^{-1}$

$$\tau = \frac{1}{(4.6 \times 10^{-2})} = 22\,s$$

Both evaluations agree.

$\tau = CR$ and so

$$C = \frac{\tau}{R} = \frac{22}{(100 \times 10^3)} = 2.2 \times 10^{-4}\,F \text{ or } 220\,\mu F$$

Uses of capacitors

Capacitors, like batteries, store electrical energy, but unlike batteries they can release this energy very rapidly. This characteristic makes the capacitor an important energy-storage component in the electric circuits used in defibrillators and camera flash guns.

Summary

- A capacitor is a circuit component that can store charge. The amount of charge it can store is called its capacitance.
- The capacitance C is given by $C = Q/V$, where Q is the charge on the capacitor when there is a potential difference V between its plates.
- The unit of capacitance is the farad (F). One farad is one coulomb per volt.
- The equivalent capacitance C of capacitors when connected in parallel is given by $C = C_1 + C_2 + C_3 + ...$
- The equivalent capacitance C of capacitors when connected in series is given by the equation:

$$\frac{1}{C} = \frac{1}{C_1} + \frac{1}{C_2} + \frac{1}{C_3} + ...$$

- The energy stored in a charged capacitor is given by the equations:

$$E = \tfrac{1}{2}QV = \tfrac{1}{2}CV^2 = \frac{Q^2}{2C}$$

- When a capacitor discharges, the charge, current and the potential difference drop off exponentially:

$$Q = Q_0\,e^{-t/CR}, \quad I = I_0\,e^{-t/CR}, \quad V = V_0\,e^{-t/CR}$$

- The time constant for the discharge is given by $\tau = CR$. This is the time for the charge (or current or p.d.) to reduce to $1/e$ or 37% of its initial value.

■ Magnetic fields

The space surrounding a magnet where a magnetic material experiences a force is called a magnetic field. We have already looked at the magnetic field associated with a bar magnet, on p. 6. The pattern of the force variation in a magnetic field is described using magnetic field lines. The direction of the magnetic field at a point is taken as the direction of the force that acts on a theoretical 'free north pole' placed at that point.

In 1819 Oersted, a Danish physicist, accidentally discovered that an electric current sets up its own magnetic field in the space surrounding the current-carrying conductor. We need to expand our definition of a magnetic field to include its generation by, and effect on, current-carrying conductors.

We will look now at the pattern of magnetic fields near differently shaped current-carrying conductors.

A magnetic field is a region of space, around a magnet or a current-carrying conductor, where a magnetic material or another current-carrying conductor experiences a force.

Magnetic field due to a straight wire

If a vertical stiff conducting wire is inserted through a hole in stiff card that is supported horizontally, an electric current of approximately 5 A passed through the conductor, and iron filings sprinkled onto the card while tapping it gently, the iron filings form themselves into a pattern of concentric circles about the wire. A small magnetic plotting compass placed on the card shows the direction of the force field and reverses when the direction of the current is reversed (Figure 18).

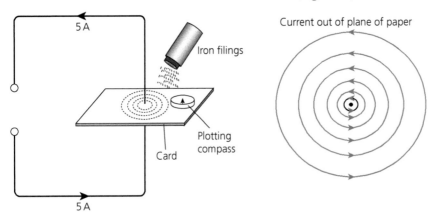

Figure 18 Arrangement to show the magnetic field due to a current-carrying wire

Exam tip

Notice that the further from the wire the field lines are, the more widely separated they are. This is to represent a weakening field.

There is a simple way to remember the direction of the field relative to the current direction; this is the **right-hand grip rule**. Imagine gripping the wire, so that your right thumb points in the direction of the current, then your fingers curl in the direction of the magnetic field lines (Figure 19).

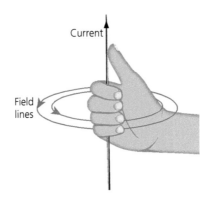

Figure 19 The right-hand grip rule

Magnetic field due to a multi-turn coil or solenoid

A solenoid is a long (cylindrical) coil with a large number of turns of wire. The magnetic field associated with a current-carrying solenoid can be investigated using iron filings and a plotting compass as before (Figure 20).

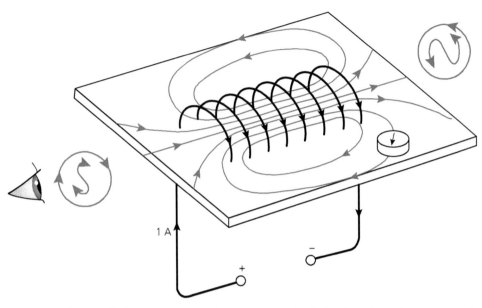

Figure 20 Arrangement to show the magnetic field of a solenoid

The magnetic field lines inside the length of the solenoid are close together, parallel and equally spaced, indicating that the field is uniform and strong. Once again the direction of the magnetic field reverses as the current direction is reversed.

The direction of the field relative to the current direction in the solenoid can be determined using the **clock rule**. When viewing one end of the solenoid, if the current is in an a**N**ticlockwise direction, that end acts like a N pole (see Figure 20); if the current is in a clockwise direction, that end acts like a S pole.

Outside the solenoid the magnetic field is the same shape as that of a bar magnet. Note that inside the solenoid the magnetic field lines go from a S pole to a N pole.

Just at the mouth of the solenoid the magnetic field lines spread out and the field strength is half that of the uniform value within the solenoid.

Force on a current-carrying conductor in a magnetic field

We have all played with small bar magnets at some stage and noted the forces of attraction and repulsion between them. There is an interaction between the two magnetic force fields. The effect of the forces may be summarised in the law of magnets.

It should therefore not be a surprise that a wire carrying a current in a magnetic field will experience a force, because the current-carrying wire has its own magnetic field, so acts as the second magnet.

This can be demonstrated with the arrangement shown in Figure 21. A wire that is free to move is placed so that part of it passes between the poles of a strong magnet. When a current is passed through the wire, the wire will move either up or down, depending on the relative directions of the magnetic field and the current.

The law of magnets states that like poles repel and unlike poles attract.

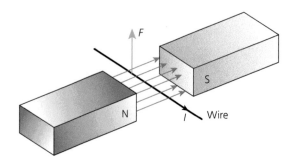

Figure 21 Force on a current carrier in a magnetic field

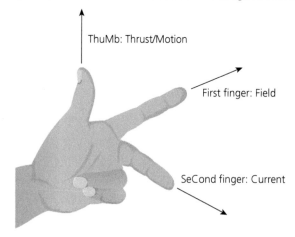

Figure 22 Fleming's left-hand rule

Exam tip

Note that the **force is always at right angles** to both the current and magnetic field. (We only need to consider the case of current carrier and field being perpendicular to one another.)

The direction of the force, causing the movement of the wire, is predicted using Fleming's left-hand rule. This is illustrated in Figure 22.

The force arises because of the interaction of the two separate magnetic fields — the uniform field between the poles of the magnet and the field due to the current in the wire. The two fields cannot exist independently in the same space; a single resultant field exists as shown in Figure 23. The field lines act as if in tension and will, if possible, 'catapult' the current carrier out or, if the current carrier is fixed, will move the external magnet (as in Figure 25).

Figure 23 The catapult effect

It is this effect that is the principle of the electric motor (Figure 24).

It is sometimes referred to as the **motor effect**.

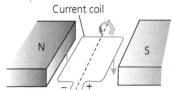

Figure 24 The electric motor principle

What affects the size of the force?

The magnitude of the force between the current-carrying conductor and the magnetic field can be investigated using the apparatus shown in Figure 25.

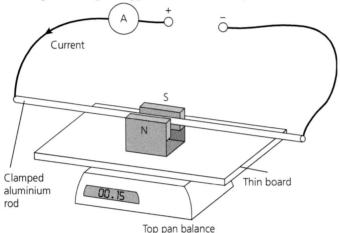

Figure 25 Arrangement to investigate the size of the magnetic force on a current-carrying conductor

The balance is set to zero after the face-pole magnets and yoke arrangement has been placed on it. A current is then passed through the fixed (non-magnetic) metal rod. Using Fleming's left-hand rule, for the situation shown, the force on the metal rod is acting upwards. Due to the interaction of the two magnetic fields there is an equal but opposite force exerted on the magnet by the field produced by the current carrier, and this force will cause the reading on the balance to increase.

If the current were reversed, a negative value would be displayed on the balance, indicating an upward, lifting force on the magnet.

The apparatus can be used to investigate how the force varies as:

1 the current is increased, by changing the voltage of the supply
2 the length of rod in the field is increased, by placing a second and third pair of magnet poles end to end on the balance
3 the strength of the magnetic field increases, by adding extra magnets to the yoke

The experiments will confirm that the force F on a current-carrying conductor placed perpendicularly to a magnetic field is directly proportional to:

■ the current I
■ the length of conductor in the field L
■ the strength of the magnetic field B

Thus we have the equation:

$$F = BIL$$

for a current-carrying conductor perpendicular to a magnetic field.

Exam tip

This can be considered to be an example of Newton's third law of action and reaction. The magnet is exerting an upward force on the current carrier so the current carrier must exert an equal and opposite force on the magnet.

Exam tip

It is important to note that if the current-carrying conductor is placed **parallel** to the magnetic field there will be **no** force.

(In general, if the conductor and field are at an angle θ to each other, then $F = BIL \sin\theta$, but you will not need to answer questions that involve this.)

Note that, unlike electrical and gravitational field strength, the symbol E is not used for the strength of the magnetic field, rather B. To compound things further, B is known as the **magnetic flux density**. This arises because the magnetic field lines are also referred to as **lines of flux**.

We can rearrange the equation $B = F/IL$, and this gives us a definition for the magnetic flux density B.

Note that, since force is a vector quantity, so is magnetic flux density: it has direction. The SI unit of magnetic flux density is the **tesla** (T).

$$1\,T \equiv 1\,N\,A^{-1}\,m^{-1}$$

The **magnetic flux density** B is the force acting per unit length on a wire carrying unit current, which is placed perpendicular to the magnetic field.

Worked example

The vertical component of the Earth's magnetic flux density is $40\,\mu T$. If the current in a horizontal overhead cable is $150\,A$, calculate the force per unit length created in the wire due to this component of the field.

Answer

The force per unit length is F/L and from $F = BIL$ this is equal to BI:

$$F/L = B \times I = (40 \times 10^{-6}) \times 150 = 6.0 \times 10^{-3}\,N$$

Note this would be a very small, negligible force.

Knowledge check 9

A stiff wire 10 cm long, of mass 4.5 g, is placed on a horizontal table. If the horizontal component of the Earth's magnetic field is 18 μT, what is the minimum current needed to pass through the wire for it to lift off the table?

Magnetic flux

Before considering the phenomenon of electromagnetic induction, we need to clarify and extend the field line model.

In a strong field the field lines are close together and the field strength or flux density is high. So the magnetic flux density can be thought of as the number of field lines, or lines of flux, that pass through unit area (Figure 26).

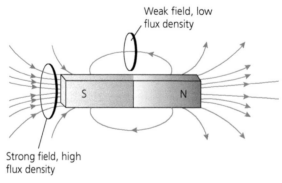

Figure 26 Flux density and field lines

The **magnetic flux**, symbol Φ, that passes through an area A, provided the area is *perpendicular* to the flux lines, is given by:

magnetic flux = flux density × area

$$\Phi = BA$$

The SI unit of flux is the weber, Wb ($1\,\text{Wb} = 1\,\text{T}\,\text{m}^2$).

Since $B = \Phi/A$, it becomes more apparent why B is referred to as flux density.

If Φ is the flux through the cross-sectional area A of a coil of N turns, the effective total flux through it, called the **flux linkage**, is $N\Phi$, since the same flux Φ links each of the N turns.

magnetic flux linkage $= N\Phi = NBA$

Knowledge check 10

A flat circular coil of mean diameter 10 cm has 25 turns. The coil is placed with its plane perpendicular to a uniform field of magnetic flux density 90 mT. Calculate the magnetic flux linkage with the coil.

Electromagnetic induction

After Oersted's discovery that a magnetic field is set up in the space surrounding a current-carrying conductor, physicists began to search for a method by which a magnetic field might induce a current to flow in a conductor. After years of frustration, success came in 1831, when Faraday in England and Henry in America announced separately that *relative motion* between a magnetic field and a conductor causes a current to flow in the conductor. Up to this point the aspect of relative motion had been missed.

When the N pole of a small permanent magnet is moving into a multi-turn coil of conducting wire connected to a sensitive centre-zero ammeter, a deflection is observed, indicating that a current is being induced in the coil (Figure 27).

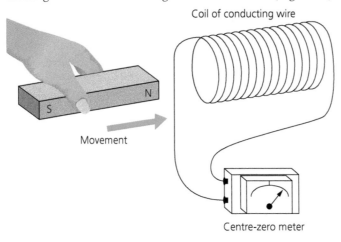

Figure 27 Simple electromagnetic induction apparatus

Note these further observations:

- A similar deflection is obtained if the coil is moved towards the N pole of the magnet.
- When the N pole is moved away from the coil, or the S pole into the coil, the deflection is in the opposite direction.
- An induced current flows only when there is relative motion between the coil and the magnet.
- The faster the relative motion, the greater the induced current.

The induced current in the coil has been caused by the creation of an **induced e.m.f.** (electromotive force) across the ends of the coil.

Faraday used the concept of lines of magnetic flux to help him picture and explain the phenomenon. The induced e.m.f. can be viewed to have been caused by either a *change* in the number of magnetic flux lines *linking* the turns of the coil, or by the wires of the coil *cutting* across the flux lines (Figure 28).

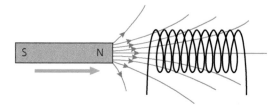

Figure 28 Flux linkage variation

Faraday's law

The e.m.f. and hence the current induced in the coil can be increased by:

- using a stronger magnet
- using a coil with more turns
- using a coil with a greater cross-sectional area
- moving the magnet faster

It can be deduced from these observations that the larger and the faster the change in the flux linkage of the coil, the greater the induced e.m.f.

This is summarised in Faraday's law.

If a change in flux $\Delta\Phi$ occurs in time Δt then:

$$\text{induced e.m.f. } E = \frac{\Delta\Phi}{\Delta t}$$

For a coil with N turns, the e.m.f. will be N times greater:

$$\text{induced e.m.f. } E = \frac{\Delta(N\Phi)}{\Delta t}$$

Lenz's law

It is observed that the direction of the induced current depends on the direction of the relative motion.

Consider the situation of a N pole of the magnet approaching a coil, as in Figure 28. The induced current in the coil will create a magnetic field around the coil, like that in Figure 20. Imagine two scenarios.

Either:

- The end of the coil nearest the magnet becomes a S magnetic pole. This would attract the magnet, accelerate the magnet and increase the rate of flux change. This in turn would increase the induced e.m.f., the induced current and the coil's magnetic field, and attract it even more. The sequence would continue and our energy worries would be over! It clearly contravenes the principle of conservation of energy, with both mechanical and magnetic energies increasing, and does not happen.

> Faraday's law states that the magnitude of the induced e.m.f. is equal to the rate of change of the flux linkage or proportional to the rate of flux cutting.

Exam tip

A stationary magnet, even if fully inserted in the coil, will not induce any current — there must be relative motion, to produce a change in flux linkage.

Or:

■ The end of the coil nearest the magnet becomes a N magnetic pole. This would repel the magnet and work would need to be done to push the magnet towards the coil. In effect there is a transfer of energy, the work done being converted into electrical energy. This is what happens.

The second idea is confirmed by considering the N pole of the magnet being removed from the coil. The induced current creates a S pole (with the current reversed, flowing clockwise) at the end of the coil, which opposes the removal of the magnet. In effect we have to pull the magnet out.

Lenz summarised this conservation of energy aspect of electromagnetic induction. **Lenz's law** states that the direction of the induced e.m.f. or current is such that it *opposes* the change producing it.

Worked example

A car is travelling at a speed of $15\,\mathrm{m\,s^{-1}}$ on a straight horizontal road in a region where the vertical component of the Earth's magnetic field is $30\,\mu\mathrm{T}$. The front of the car's roof-rack is a metal cross-bar. Due to the motion in the magnetic field, an e.m.f. of $0.5\,\mathrm{mV}$ is induced between the ends of the cross-bar. Calculate the length L of the cross-bar.

Answer

The cross-bar will move through the field, cutting the flux lines (Figure 29).

Figure 29 Plan view

The area swept out per second will be the product of the length L and the speed $v = \Delta s/\Delta t$ where s is the distance travelled:

$$E = \frac{\Delta(N\Phi)}{\Delta t} = \frac{\Delta(NBA)}{\Delta t} = \frac{NB\Delta(Ls)}{\Delta t} = \frac{NBL\Delta s}{\Delta t} = NBLv$$

$$E = NBLv$$

$$L = \frac{E}{NBv}$$

$$L = \frac{(0.5 \times 10^{-3})}{1 \times (30 \times 10^{-6}) \times 15} = 1.1\,\mathrm{m}$$

A.C. generator

A generator produces electrical energy by electromagnetic induction. In its simplest form, it consists of a coil which is rotated between the poles of a magnet so that the flux linkage is continuously changing (Figure 30a). The flux Φ linking each turn of a coil, of area A, in a uniform field B, varies sinusoidally with time t (Figure 30b), and is given by the equation:

$$\Phi = BA \cos \omega t$$

where ω is the angular velocity of the rotation and $\omega = 2\pi f$.

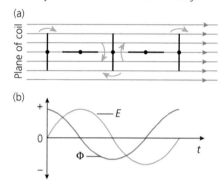

Figure 30 Principle of a generator

The e.m.f. E induced across the ends of the coil, with N turns, will as a result also vary sinusoidally (also shown in Figure 30b), and is given by the equation:

$$E = BAN\omega \sin \omega t$$

where ω is the angular velocity of the rotation.

As the maximum value of sin is 1, the maximum or peak e.m.f. is

$$E_0 = BAN\omega$$

This expression shows that the e.m.f. increases with B, A, N and ω.

The sinusoidally varying e.m.f. produced will generate **a.c.**, alternating current. For a simple single-coil a.c. generator the frequency of the a.c. generated equals the number of revolutions per second of the coil.

Worked example

The output voltage E of a simple a.c. generator changes with time t (in seconds) as given by:

$$E = 300 \sin (314t)$$

a Calculate the frequency of the a.c. generated.

b Calculate the minimum time for the output of the generator to rise from zero to 150 V in each cycle of operation.

Answer

a $\omega = 314\,\text{s}^{-1} = 2\pi f$

   frequency $f = 50\,\text{Hz}$

b $E = 300\sin(314t) = 150$

   $\sin(314t) = 150/300 = \tfrac{1}{2}$

   $314t = \sin^{-1}\tfrac{1}{2} = \pi/6$

   $t = 1.66 \times 10^{-3}\,\text{s}$ or $1.66\,\text{ms}$

Knowledge check 12

A simple generator has a 300-turn rectangular coil of dimensions 20 mm × 35 mm. The coil rotates in a uniform magnetic field of flux density 0.25 T. How many revolutions must the coil make per second in order to produce a peak output of 12V?

Transformers

A **transformer** is another device based on the principle of electromagnetic induction. A transformer is used to change the value of an **alternating** voltage.

It consists of two coils of insulated wire, called the primary and the secondary, wound around an iron core, either one on top of the other as in Figure 31a or on separate limbs of the core, as in Figure 31b. There is no electrical connection between the coils.

Exam tip

The presence of the iron ensures that all the flux associated with one coil passes through the other. It is referred to as 'soft iron' because it is easily magnetised and demagnetised.

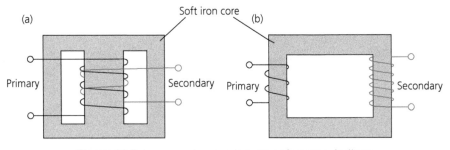

Figure 31 Primary and secondary transformer windings

When an alternating potential difference is applied across the primary, the resulting current produces an alternating magnetic flux which links the secondary. This means that the flux linkage in the secondary is continually changing and the conditions for an induced e.m.f. exist. As a result an alternating e.m.f. is induced across the secondary.

It can be shown that, to a good approximation:

$$\frac{\text{e.m.f. induced in secondary}}{\text{p.d. applied to primary}} = \frac{\text{number of secondary turns}}{\text{number of primary turns}}$$

So, for a transformer with N_P and N_S turns on the primary and secondary respectively and with e.m.fs across the coils V_P and V_S:

$$\frac{V_S}{V_P} = \frac{N_S}{N_P}$$

When V_S is larger than V_P, there will be more turns on the secondary coil than the primary and the transformer is referred to as a 'step-up transformer'.

Exam tip

When Faraday originally discovered electromagnetic induction, his apparatus consisted of two coils and no magnet. Increasing the current in a coil is equivalent to bringing a magnet closer, and reversing the direction of current is equivalent to reversing the polarity of the magnet or the direction of the field.

Content Guidance

Likewise, when V_S is smaller than V_P, there will be fewer turns on the secondary coil than the primary and the transformer is referred to as a 'step-down transformer'.

If the p.d. is stepped up by a transformer, the current is stepped down, in the same ratio if the transformer is 100% efficient. This follows from the conservation of energy:

power into primary = power out of secondary

$$V_P I_P = V_S I_S$$

Therefore:

$$\frac{V_S}{V_P} = \frac{I_P}{I_S} = \frac{N_S}{N_P}$$

This is the case for an 'ideal' transformer. In practice, while many transformers have very high efficiency there are always some power losses. Table 3 lists the reasons for power losses and how these can be minimised.

Table 3 Power losses in transformers

Power loss due to	Minimised by
Leakage of magnetic flux between primary and secondary	Common-core winding, as in Figure 31a
Resistive heating of the wire in the primary and secondary coils	Use of thick wire of low-resistivity material
Heating within the core due to induced currents in the core called eddy currents	Laminating the core
Hysteresis, which is heating in the core due to magnetic field reversal	Use of soft iron

Worked example

A transformer is used with a mains 240 V a.c. supply to power a lamp rated at 12 V, 24 W. The secondary coil of the transformer has 60 turns. Assume that the transformer is ideal.

a What type of transformer is this?
b Calculate the number of turns on the primary of this transformer.
c What is the current drawn from the mains supply?

Answer

a Step-down

b $\frac{V_S}{V_P} = \frac{N_S}{N_P}$

$$\frac{12}{240} = \frac{60}{N_P}$$

$N_P = 1200$ turns

c Using $P = VI$, secondary current $I_S = 24/12 = 2\,A$

$$\frac{V_S}{V_P} = \frac{I_P}{I_S}$$

$$\frac{12}{240} = \frac{I_P}{2}$$

$$I_P = 0.1\,A$$

Transmission of electrical energy in the national grid

The national grid is the network of transmission lines which connect power stations to the electricity consumers. Since power is the product of voltage and current, a given amount of power can be transmitted either at high voltage and low current, or at low voltage and high current. The cables that transmit the electrical energy have resistance, and therefore some of the energy is wasted by producing heat in the cables as the current passes through them.

If the resistance of the cables is R, the heat energy produced in time t when a current I is flowing through them is I^2Rt. The amount of energy that is wasted is thus proportional to the square of the current in the cable. Therefore the most effective way to transmit power is at a high voltage with low current.

Another advantage of high-voltage transmission is that lower current requires thinner and therefore cheaper and lighter cable. The disadvantage is the high cost of isolating the high-potential cables high above the ground on pylons (or deep below ground in highly protective casings).

Modern power station generators produce alternating current at a frequency of 50 Hz at about 25 kV. This is then stepped up using transformers to 132 kV, 275 kV or 400 kV before transmission. It is later stepped down by other transformers to 33 kV (for heavy industry), 11 kV (for light industry), or 230–240 V (for domestic use).

Kilroot power station supplies 520 MW of electricity to the Northern Ireland grid at 275 kV, which corresponds to a transmission line current of 1890 A. Multi-strand aluminium cables reinforced with steel are used in the transmission lines.

Exam tip

The national grid uses a.c. rather than d.c. (direct current) because the most efficient way to convert low voltages to high voltages and vice versa is by using transformers, and these cannot operate on d.c.

Knowledge check 13

A small power station generates 20 MW of electricity. How much power is lost in cables of resistance 4 Ω if the electricity is transmitted at (a) 5 kV, (b) 50 kV?

Summary

- A magnetic field is a region of space where a force is exerted on magnetic materials and current-carrying conductors.
- A current-carrying conductor produces its own magnetic field, the strength and direction of which depend on the size and direction of the current and the shape of the conductor.
- There is a force on a current-carrying conductor whenever it is at an angle to a magnetic field.
- The magnitude of the force on a conductor of length L carrying current I, placed at right angles to a magnetic field of flux density B is given by $F = BIL$.
- The magnetic flux density B is numerically equal to the force acting per unit length, on a wire carrying unit current, when at right angles to the magnetic field.
- The magnetic flux Φ is the product of magnetic flux density and area perpendicular to the flux: $\Phi = BA$.

- For a coil with N turns the flux linkage equals NBA.
- Faraday's law tells us that the size of the induced e.m.f. is equal to the rate of change of flux linkage.
- Lenz's law tells us that the direction of the induced e.m.f. is such as to oppose the change producing it.
- Faraday's and Lenz's laws are summarised by the equation:

$$E = \frac{-\Delta(N\Phi)}{\Delta t}$$

- The induced e.m.f. from a simple a.c. generator is given by:

$$E = BAN\omega \sin \omega t$$

- For an ideal transformer:

$$\frac{N_S}{N_P} = \frac{V_S}{V_P} = \frac{I_P}{I_S}$$

Deflection of charged particles in electric and magnetic fields

Electric field deflection

It follows from the definition of electric field strength (p. 14) that a charge q in a field of strength E is subject to a force F, given by $F = qE$, and has an acceleration $a = F/m = qE/m$.

For a uniform electric field we noted that the field strength is the same at all points, $E = V/d$, therefore the force on a charged particle is the same no matter where it is placed in the field.

If charged particles enter a uniform electric field that acts at an angle to their direction of motion, they are deflected from their original path (Figure 32), much in the same way as a mass projected in a gravitational field.

The direction of force on the positive charge in Figure 32 is the direction of the electric field, so is constant and independent of the charged particle's velocity. This results in the particle moving in a **parabolic** path. The force on a negative charge would be in the opposite direction so in the electric field of Figure 32 its path would curve upwards towards the positive plate.

Exam tip

The charged particles could enter at any angle, as with any projectile, not just in a direction perpendicular to the field as in the example shown in Figure 32.

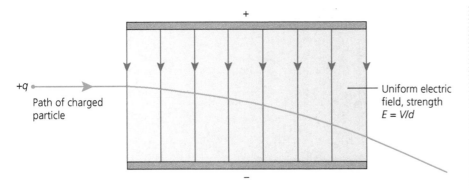

Figure 32 Charged particle deflection in an electric field

For a charged particle moving *along* an electric field line, the motion is simpler. The particle will accelerate (if it is a positive charge) in a straight line, like a mass dropped towards the surface of the Earth.

Magnetic field deflection

An electric current is a movement of charge and so the force experienced by a current-carrying conductor in a magnetic field is in effect the resultant of the forces acting on the moving charges. Applying this logic we can adapt the equation for the force on a current-carrying conductor to find the force on a single moving charge in a magnetic field.

Consider a length of conductor L carrying N charged particles, each of charge q, taking time t to travel length L. The force on the conductor is $F = BIL$ where the magnetic field is uniform and perpendicular to the current. But $I = Nq/t$. Therefore:

$$F = \frac{B(Nq)L}{t}$$

Substituting speed $v = L/t$:

$$F = B(Nq)v$$

Hence the force on *one* charged particle is:

$$F = Bqv$$

if the charge is moving at right angles to the field.

The direction of the force is given, as before, by using Fleming's left-hand rule (p. 30), remembering that conventional current direction is the direction of *positive* charge movement. The force is **always at right angles to both the field and the direction of charge movement**. This is a key difference from the force in an electric field and is responsible for the difference in path.

Since the deflecting force is always perpendicular to the path of the charge movement, it only changes the direction of motion and not the speed — there is no force component parallel to the motion to cause linear acceleration. This is the condition seen in circular motion — the force is centripetal. Consequently when charged particles enter a uniform magnetic field at right angles they are deflected into a circular path (Figure 33).

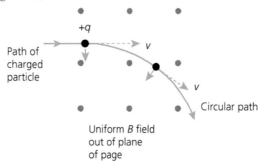

Figure 33 Charged particle deflection in a magnetic field

Exam tip

When charges are moving parallel to a magnetic field they experience no force, just as no force is created when a current-carrying conductor is parallel to a magnetic field.

Worked example

A beam of electrons, moving with a speed of $4.0 \times 10^6 \, \text{m s}^{-1}$, enter a uniform magnetic field at right angles. The field has a flux density of 1 mT. Determine the radius of the circular path followed by the electrons.

Answer

The force on the moving charge in the magnetic field is the centripetal force:

$$F = Bqv = \frac{mv^2}{r}$$

Therefore:

$$r = \frac{mv}{Bq} = \frac{m_e v_e}{Be}$$

because the charges are electrons. So

$$r = \frac{(9.1 \times 10^{-31}) \times (4.0 \times 10^6)}{(1 \times 10^{-3}) \times (1.6 \times 10^{-19})}$$

$$r = 2.3 \times 10^{-2} \, \text{m or } 2.3 \, \text{cm}$$

Summary

- A charge q in a uniform electric field will experience a constant force $F = qE$.
- The direction of the force is the direction of E if the charge is positive, or the opposite direction if the charge is negative.
- When a charged particle is projected into a uniform electric field at an angle it will follow a parabolic path.

- The force on a charge q moving perpendicularly to a uniform magnetic field is given by $F = Bqv$.
- The force is always perpendicular to both the magnetic field and the direction of charge movement, given by Fleming's left-hand rule.
- When a charged particle is projected perpendicular to a uniform magnetic field it will follow a circular path.

Particle accelerators

As we have seen, charged particles can be accelerated by electric and magnetic fields. But what are the reasons for accelerating these subatomic particles?

Cathode rays

One of the simplest charged particle accelerators is the 'electron gun'. The electron gun is an electrode assembly for producing a narrow beam of fast-moving electrons. The gun comprises a heated cathode and one or more anodes with circular holes. The resulting beam was originally called a 'cathode ray', before it was known that it was composed of electrons.

The electron gun of a cathode ray tube is used in cathode ray oscilloscopes and in X-ray tubes, and was used in early televisions (Figure 34). In an evacuated vessel, electrons are emitted from a cathode heated by a wire filament. The electrons experience a force in the electric field between the cathode and the anode, and as a result are accelerated to very high speeds.

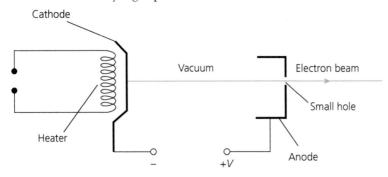

Figure 34 Basic structure of an electron gun

The potential difference V between cathode and anode can be used to calculate the velocity change of the electrons. The volt is defined as a joule per coulomb, which means that a charge of 1 C accelerated through a potential difference of 1 V will gain 1 J of energy. Effectively:

work done in electric field = loss of electrical potential = gain in kinetic energy

$$W = qV = \tfrac{1}{2}mv^2$$

In an X-ray tube electrons emitted from a hot cathode in an evacuated tube are accelerated by a potential difference of 1000 V. Calculate the speed acquired by these electrons.

Answer

Using $W = qV = \tfrac{1}{2}mv^2$:

$$eV = \tfrac{1}{2}m_e v_e^{\,2}$$

→

$$v_e^2 = \frac{2eV}{m_e} = \frac{2 \times (1.6 \times 10^{-19}) \times 1000}{(9.1 \times 10^{-31})} = 3.5 \times 10^{14}\,\mathrm{m^2 s^{-2}}$$

$$v_e = 1.9 \times 10^7\,\mathrm{m\,s^{-1}}$$

Nuclear artillery

A second, less practical use of accelerated charged particles is to provide the 'artillery' for physicists to probe the very heart of matter and discover its secrets.

Early nuclear physics experiments were carried out using alpha particles, emitted by radioactive substances, as the 'bullet' to fire at matter. Alpha particles from radioactive sources have limitations, due partly to the fact that their energy never exceeds 10 MeV, and to the difficulty in obtaining and handling sufficiently strong sources. It was in the early 1930s that Cockcroft and Walton first used artificially accelerated particles (protons) in a collision-type experiment to confirm Einstein's $E = mc^2$ equation.

Particle accelerators are capable of generating large numbers of subatomic particles with a wide range of energies. Beam currents of a few micro-amperes are possible, equivalent to about 10^{13} particles per second. Since the accelerators produce particles travelling in just one direction, the number hitting a target may be about 1 million times greater than that for a radioactive source. Additionally, the ability to accelerate a variety of different particles, for example electrons and protons, rather than just alpha particles, permits a greater range of experiments to be performed.

The particle energy achieved by the latest accelerators can go as high as 1 TeV. This enables the colliding particles to penetrate further, and consequently attention has turned from the structure of the atom to that of the nucleus.

There is a variety of modern accelerator designs (Figure 35).

Knowledge check 15

Calculate the final speed of a proton after it has been accelerated by a potential difference of 1000 V. Express the speed as a percentage of the speed of light.

Exam tip

Accelerated particles may also be used for bombarding nuclei to produce short-lived radioisotopes for medical use.

Exam tip

The electronvolt (eV) is a small unit of energy. It is equal to the energy gained by an electron in being accelerated by a potential difference of 1 volt. 1 eV is equivalent to 1.6×10^{-19} J. It is a much more convenient unit of energy than the joule when dealing with nuclear physics.

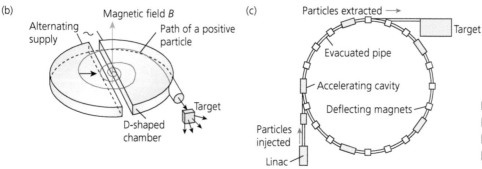

Figure 35
(a) Linear accelerator,
(b) cyclotron,
(c) synchrotron

- Linac: linear accelerator, can be up to 3 km long, can accelerate electrons to about 50 GeV
- Cyclotron: used to accelerate heavy particles such as protons and alpha particles to energies of about 25 MeV
- Synchrotron: combines the technology of the linac and cyclotron, can accelerate protons to 1 TeV

The general principles are the same:

- The particles must be charged, otherwise they will not experience a force in electric or magnetic fields.
- Electric fields exert a force on the charged particles and accelerate them, thereby increasing their energy. It is not possible to maintain a potential difference greater than 200 kV between two electrodes in an evacuated tube, so for acceleration to energies greater than 200 keV multi-stage accelerators are used.
- Magnetic fields change the direction of the moving charges so that they move in a circular path (otherwise the length of the accelerator would be too great) and can continue to be accelerated during many orbits.

Synchrotron

In a synchrotron the injected charged particles are kept in an orbit of fixed radius as they are accelerated. This is achieved using by using a magnetic field of increasing strength at intervals over the circumference of the orbit. The radius of the orbit is given by the equation $r = mv/Bq$ (see the worked example on p. 42).

A radio-frequency alternating p.d. is applied to accelerating 'cavities' in the particles' path (Figure 35) so the particles are attracted as they approach and repelled as they leave. As the particles accelerate they take less time to complete their orbit, therefore the frequency of the alternating p.d. applied to the cavities must also increase so that the electric field remains synchronised with the particles as they pass through.

Particles injected into the synchrotron have usually been pre-accelerated in a linac, which is positioned tangentially to the ring.

In addition to the magnetic fields that bend the beam into a circular path, magnetic fields are used to focus the particle beam, and to extract the accelerated beam out of the ring. Several extraction points may exist in the ring which will allow the same machine to be used for a number of different experiments simultaneously. Alternatively, some synchrotrons have two loops which overlap to collide particles and antiparticles head on.

Electromagnetic radiation, referred to as 'synchrotron radiation', is released by a charged particle being accelerated in a circular path. The radiation loss is greater the lower the mass of the charged particles and the smaller the radius of orbit.

The upward sweep of the values of the magnetic field strength and the accelerating p.d. frequency usually takes about one second and is repeated every few seconds. At each sweep a bundle of particles is accelerated to full energy. During this short time the particles execute many million revolutions at speeds approaching the speed of light and may travel hundreds of thousands of kilometres.

Exam tip

At speeds approaching the speed of light, relativistic effects mean that the energy supplied will increase the particle's mass rather than its speed. In an electron synchrotron the electrons are already travelling close to the speed of light when injected which leads to their orbital period remaining constant and so a constant a.c. frequency can be used.

Worked example

What is the time for one revolution of protons in a synchrotron when a magnetic field of 0.8 T is applied at the deflecting magnets?

Answer

Equating the magnetic force to the centripetal force,

$$Bqv = \frac{mv^2}{r}$$

Therefore:

$$v = \frac{Bqr}{m}$$

The time for the particle to make one revolution is the period, $T = 2\pi r/v$.

Therefore:

$$T = \frac{2\pi m}{Bq}$$

$$T = \frac{2\pi \times (1.67 \times 10^{-27})}{0.8 \times (1.6 \times 10^{-19})} = 8.2 \times 10^{-8}\,\text{s}$$

(Frequency of rotation $f = 1/T = 1.2 \times 10^7\,\text{Hz}$)

Exam tip

The frequency of the alternating p.d. has to be greater than the frequency of rotation, because it is reversing polarity during the time the beam spends outside the cavity, and there is more than one accelerating cavity in the ring.

The Super Proton Synchrotron (SPS) at CERN in Geneva has been in operation since 1976. Protons from a 50 MeV linear accelerator are boosted to 10 GeV in a small synchrotron and then injected into the 2.2 km diameter main ring of the SPS. The beam's circular path is maintained by more than 700 deflecting magnets, each with a field of up to 1.8 T. There are also over 200 focusing magnets positioned around the ring. Two radio-frequency cavities accelerate the beam. Beam pulses of the order of 10^{13} protons can be produced, with energy up to 450 GeV.

Relativistic mass

Consider a proton, accelerated so that it gains energy 450 GeV.
Using $m_p = 1.67 \times 10^{-27}$ kg, calculate the speed of the accelerated protons.

$$\Delta E = \tfrac{1}{2}mv^2$$

$$(450 \times 10^9) \times (1.6 \times 10^{-19}) = \tfrac{1}{2} \times (1.67 \times 10^{-27}) \times v^2$$

$$v = 9.3 \times 10^9\,\text{m s}^{-1}$$

But note this is well in excess of c, the speed of light, $3.0 \times 10^8\,\text{m s}^{-1}$, and according to Einstein's special theory of relativity nothing can travel beyond the speed of light!

Einstein's solution to the problem was that, as work is done on the particles, it is still transferred as kinetic energy, however the velocity does not increase as much as we would expect because there is an increase in the *mass* of the particle.

Relativistic mass increase becomes significant above an energy value of 25 MeV, and this can create challenges in accelerator designs. Synchrotrons overcome this by changing the magnetic field strength to take account of the changes in both mass and speed to maintain an orbit of fixed radius.

Antimatter

In 1932 Anderson examined traces created by cosmic ray collisions in a cloud chamber and discovered a new fundamental particle — a positively charged electron. That is, it had the same mass and magnitude of charge as an electron, but a positive rather than a negative charge. The positive electron was named a positron. The discovery of the positron revived interest in an earlier proposal published by Dirac, that for every subatomic particle there should exist an **antimatter** particle or **antiparticle** — a particle of the same mass and carrying charge of the same magnitude but of opposite sign.

Before long, other antiparticles were found. Antiprotons were first observed in 1955 in the collisions of an accelerator experiment. They were produced when copper nuclei were bombarded by protons with energy greater than 5.6 GeV.

Antiparticles bind with one another to form antimatter, just as ordinary particles bind to form normal matter. For example, a positron and an antiproton can form an antihydrogen atom. In 1995 at CERN, nine atoms of antihydrogen were created, each lasting 40 nanoseconds.

Annihilation

Antimatter particles are difficult to find because the observable universe is composed almost entirely of ordinary matter; whenever antimatter particles appear they quickly meet up with their equivalent ordinary matter particle and they annihilate one another.

In **annihilation** the combined mass of the particle and antiparticle is converted into energy. For example, when an electron and a positron collide they annihilate and produce two gamma-ray photons (Figure 36):

$$_{-1}^{0}e^- + _{+1}^{0}e^+ \rightarrow 2_0^0\gamma$$

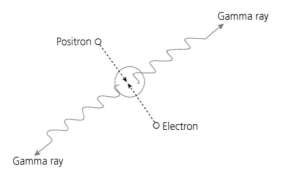

Figure 36 Electron–positron annihilation

> **Exam tip**
>
> Rest mass m_0 is constant, but the mass m of a particle in motion increases with velocity according to the equation:
>
> $$m = \frac{m_0}{\sqrt{1 - \left(\frac{v}{c}\right)^2}}$$
>
> When $v = 0.10c$, $m \approx m_0$. However, when $v = 0.99c$, $m \approx 7m_0$.

> **Exam tip**
>
> Some antiparticles have special names and symbols but most are represented by a bar over the particle symbol. For example, an antiproton is represented by $\bar{p}$.

> **Exam tip**
>
> Uncharged particles such as neutrons and neutrinos have antiparticles that differ from the particle in other ways, such as quark composition or lepton number.

The important conservation principles of classical physics also apply to nuclear physics, namely:

- conservation of mass-energy
- conservation of charge
- conservation of momentum
- conservation of nucleon number (this has emerged from nuclear physics itself)

A single photon of the correct energy could satisfy three of the conservation requirements, but not momentum conservation. (In 1923 Compton had shown that photons, although massless, carry momentum.) Consider the case of a head-on collision of a slow-moving electron and positron. The total momentum before the collision is zero. For zero momentum after the collision we need to have two identical photons moving in opposite directions. The two-photon explanation tallied with observations already recorded in 1930 by a physicist called Chao.

Worked example

An electron and a positron both with negligible kinetic energy annihilate, producing two identical photons. Calculate:

a the energy of the photons, in eV

b the wavelength of the photons

Answer

a $E = \Delta mc^2 = (2 \times 9.11 \times 10^{-31}) \times (3.00 \times 10^8)^2 = 1.64 \times 10^{-13}\,\text{J}$

$$1.64 \times 10^{-13}\,\text{J} = \frac{(1.64 \times 10^{-13})}{(1.6 \times 10^{-19})} = 1.02 \times 10^6\,\text{eV}$$

As there are two photons, the energy of each is 0.51 MeV.

b Using $E = hf = hc/\lambda$:

$$\lambda = \frac{hc}{E} = \frac{(6.63 \times 10^{-34}) \times (3 \times 10^8)}{(0.82 \times 10^{-13})} = 2.42 \times 10^{-12}\,\text{m}$$

This is in the gamma-ray wavelength band.

Pair production is the opposite process to annihilation. *In isolation* it would require a pair of photons meeting to produce an electron–positron pair. However, the chance of two photons meeting in this way is negligible, and what actually happens is that a single high-energy photon passes close to a nucleus, which recoils as the particle–antiparticle pair is produced. In this way energy and momentum are conserved. (Pair annihilation to a single photon can also occur near a third particle.)

When sufficient energy is available, annihilation of a particle–antiparticle pair can produce short-lived particles. The gamma energy from the annihilation is converted back into matter. This is one way in which new particles are created in accelerator experiments. In 1956 Cork discovered the antineutron in the aftermath of proton–antiproton collisions at the Lawrence Berkeley National Laboratory, just one year after the antiproton was discovered.

Summary

- Synchrotrons were developed to boost particles to much higher energies than those produced by cyclotrons.
- In the synchrotron both the field strength of the deflecting magnets and the radio-frequency of the alternating p.d. applied to the accelerating cavities are increased as the particles gain speed.
- Particles are accelerated to speeds approaching the speed of light, therefore relativistic mass changes need to be considered.
- All particles have antimatter counterparts. A particle and its antiparticle have the same mass, but they carry equal but opposite charge.
- Particles and antiparticles may annihilate if they come together, with the creation of two photons, thereby obeying the laws of conservation of energy and momentum.

Fundamental particles

We may consider atoms as the building block of matter, but we now know that an atom has electrons outside a nucleus containing protons and neutrons. But these in turn are made up of smaller particles. As a result of cosmic ray and accelerator collision experiments, more than 200 different subatomic particles have been discovered.

Currently it is thought that there are three families of fundamental particles: leptons and quarks that make up matter, and the 'exchange particles' called gauge bosons.

Fundamental particles, also known as elementary particles, are particles that cannot be broken down into any further constituents, i.e. have no internal structure.

Fundamental forces

So we think the building blocks of matter are leptons and quarks, but what holds these together to make everything exist as you see around you? The electric, magnetic and gravitational forces we have considered so far cannot explain how small, like-charged particles can be held close together.

It is thought that all the interactions in the universe are the result of four fundamental forces:

- **Gravitational**. This force between masses has an infinite range and is always attractive. It is the weakest of the four forces and while it can be ignored inside atoms, it is the obvious dominating force in our everyday lives due to the presence of one large mass, the Earth.
- **Electromagnetic**. In the 19th century, Maxwell proved that electric and magnetic forces were in effect the same type of force — the electromagnetic force. This force affects charged particles, producing the interatomic and intermolecular forces of attraction and repulsion that hold atoms and molecules together. It is responsible for the chemical, mechanical and electrical properties of matter. This also has an infinite range.
- **Weak nuclear force**. This force acts on all particles, both leptons and quarks. Its range does not extend beyond the nucleus. This is the force responsible for β⁻ decay, when a neutron decays to a proton and creates an electron and antineutrino in the process. (The antineutrino is not affected by the other forces.)

- **Strong nuclear force**. Leptons do not experience the strong force. This is the strongest force, but it has a very short range. It acts between neighbouring nucleons, holding them together. It must be strongly attractive to overcome the electromagnetic repulsion force between protons in a nucleus, but, at shorter range, it must be repulsive to prevent the nucleus from imploding. For internucleon separations of about 1 to 3 fm the strong force is attractive but for separations less than 1 fm it is repulsive.

Recent particle experiments suggest that the electromagnetic and the weak nuclear force are of the same type, referred to as the 'electroweak' force. So the four forces could be considered as three, and perhaps eventually there may be unification of all into a single force type.

We know that forces exist — in some cases we can feel them and in others we can observe their effect — but how is it that they arise? The explanation is that force is due to certain 'virtual particles', called **gauge bosons**, being exchanged between the interacting bodies. The gauge bosons are acting as messengers that convey the force, and can give rise to both attractive and repulsive forces. They are referred to as 'virtual' because they are very short lived and exist only within the context of their interaction. Different gauge bosons act as **exchange particles** for each type of force (see Table 4). All have been discovered apart from the graviton.

Table 4 The four fundamental forces

Force	Acts on	Comparative strength	Range	Exchange particle (gauge boson)
Strong nuclear	Quarks	1	10^{-15} m	Gluon
Electromagnetic	Charged particles	10^{-2}	∞	Virtual photon
Weak nuclear	Quarks and leptons	10^{-5}	10^{-17} m	Z^0, W^+, W^-
Gravitational	Particles with mass	10^{-40}	∞	Graviton

Classification of fundamental particles

We have already classified the gauge bosons; while these are fundamental particles, they do not make up the matter that we observe in our universe. All ordinary matter is made up of **leptons** and **quarks**.

Leptons

There are 12 different members of the lepton family, in three 'generations' (see Table 5). The three generations of 'electron' are the (ordinary) electron (e), the heavy electron (muon) and the superheavy electron (tau). Each generation electron has an antiparticle and all have corresponding neutrinos and antineutrinos. Only the first generation electron, e, occurs in ordinary matter. There are three important facts about leptons:

- Leptons are not affected by the strong force.
- Leptons have no internal structure and are therefore fundamental.
- Each lepton is given a **lepton number**, L, of 1; their antiparticle has L = −1. The lepton number is conserved during any reaction between particles.

Leptons are fundamental particles. The electron is the lepton that is evident in ordinary matter.

Table 5 The three generations of leptons

Lepton	Symbol	Relative charge/e	Relative mass/m_e	Lepton number	Antiparticle
Electron	e^-	−1	1	+1	positron e^+
Electron neutrino	v_e	0	0	+1	$\overline{v_e}$
Muon	μ^-	−1	207	+1	mu-plus μ^+
Muon neutrino	v_μ	0	0	+1	$\overline{v_\mu}$
Tau	τ^-	−1	3500	+1	tau-plus τ^+
Tau neutrino	v_τ	0	0	+1	$\overline{v_\tau}$

Exam tip

The neutrino is (along with the photon) the lightest of all particles and was predicted in 1933 to account for unexplained energy variation seen in beta decays and to uphold the conservation of energy. Conservation of charge requires the neutrino to have zero charge.

Quarks

The strong nuclear force divides particles that make up matter into two classes — those that it does not affect, the leptons, and those that it does affect, called **hadrons**. Nucleons, for example, are hadrons. Hadron is from the Greek word *hadros*, meaning bulky. Lepton comes from *leptos*, meaning small.

With the development of ever more energetic accelerators, over 100 different hadrons were discovered by the early 1960s, and there was no obvious pattern between the particles. Then a scattering experiment carried out at Stanford revealed structure inside the hadrons. The accelerated electrons (6 GeV) had sufficiently high energy to probe nucleons. The results confirmed a classification theory that had been put forward by Gell-Mann and Zweig independently.

They suggested that hadrons are not fundamental particles; rather hadrons are composed of smaller constituents, named **quarks** by Gell-Mann. Using the quark model it is possible to describe all hadrons in terms of combinations of different quarks and antiquarks.

There are two distinct groups of hadrons — the **baryons** and the **mesons**:

- Baryons are combinations of three quarks, not necessarily of the same type.
- Antibaryons are combinations of three antiquarks, not necessarily of the same type.
- Mesons (and their antiparticles) are composed of a quark and an antiquark, again not necessarily of the same type.

Quarks are believed to be fundamental particles. There are six types or 'flavours' of quark (see Table 6): down (d), up (u), strange (s), charm (c), bottom (b) and top (t), along with their corresponding antiparticles, the antiquarks. The d, s and b carry a charge of $-\frac{1}{3}e$, while u, c and t carry $+\frac{2}{3}e$. The corresponding antiquarks have a charge of the same magnitude but opposite sign.

The total charge of the hadron is the sum of the constituent quark charges. Quarks combine together to make particles with a total relative charge of 0 or ±1. (Relative charge is charge value relative to $+e$.)

Each quark is assigned a **baryon number** of $+\frac{1}{3}$ and the antiquark $-\frac{1}{3}$. This gives baryons an overall baryon number of 1 and their antiparticles −1. Mesons and their antiparticles have baryon number 0, as they are not baryons. (See Table 7.) The baryon number is useful in clarifying if a reaction is possible or not, because baryon number, like charge and lepton number, is subject to a conservation rule.

Hadrons are not fundamental particles. They are composed of quarks — either three quarks, forming a baryon, or a quark and antiquark, forming a meson. Neutrons and protons are examples of baryons.

Particle reactions obey the following conservation rules:

- charge is conserved
- lepton number is conserved
- baryon number is conserved

This mean the numbers must balance on each side of a reaction equation.

Table 6 Quarks

Quark	Symbol	Relative charge/e	Baryon number
Down	d	$-\frac{1}{3}$	$+\frac{1}{3}$
Up	u	$+\frac{2}{3}$	$+\frac{1}{3}$
Strange	s	$-\frac{1}{3}$	$+\frac{1}{3}$
Charm	c	$+\frac{2}{3}$	$+\frac{1}{3}$
Bottom	b	$-\frac{1}{3}$	$+\frac{1}{3}$
Top	t	$+\frac{2}{3}$	$+\frac{1}{3}$

Only two types of quark, the up and down, are needed to account for the properties of the neutrons and protons which together make up almost all of everyday, observable matter. The two antiquarks $\bar{u}$ and $\bar{d}$ are needed to explain the existence of the antiproton and the antineutron. The antineutron, like the neutron, has no overall charge, however the charges on its three antiquarks are each the opposite of those on the quarks in the neutron: neutron u ($+\frac{2}{3}$), d ($-\frac{1}{3}$), d ($-\frac{1}{3}$); antineutron $\bar{u}$ ($-\frac{2}{3}$), $\bar{d}$ ($+\frac{1}{3}$), $\bar{d}$ ($+\frac{1}{3}$).

Table 7 Quark make-up of some baryons, antibaryons and mesons

Baryon/antibaryon	Quark structure	Baryon number	Meson	Quark structure	Baryon number
Proton p	uud	+1	Pion π^+	$u\bar{d}$	0
Neutron n	udd	+1	Pion π^-	$\bar{u}d$	0
Antiproton $\bar{p}$	$\bar{u}\bar{u}\bar{d}$	−1	Pion π^0	$u\bar{u}$ or $d\bar{d}$	0
Antineutron $\bar{n}$	$\bar{u}\bar{d}\bar{d}$	−1			

Exam tip

The only stable baryon is the proton. All baryons ultimately decay into protons. There are no stable mesons. They rapidly decay into leptons and photons (energy).

Worked example

a Use the conservation of charge, lepton number and baryon number to decide if the following reactions are possible:

i electron capture:

$$p + e^- \rightarrow n + \bar{v}_e$$

ii muon decay:

$$\mu^- \rightarrow e^- + \bar{v}_e + v_\mu$$

b Identify the unknown particle U in the equation for β^- (electron) decay:

$$n \rightarrow p + e^- + U$$

Answer

a i $p + e^- \rightarrow n + \bar{v}_e$

Charge: +1 + (−1) → 0 + 0 ✓

Baryons: +1 + 0 → +1 + 0 ✓

Leptons: 0 + 1 → 0 + (−1) ✗

This reaction is not possible. (The electron antineutrino should be an electron neutrino.)

ii $\mu^- \rightarrow e^- + \overline{\nu}_e + \nu_\mu$

Charge: $-1 \rightarrow -1 + 0 + 0$ ✓

Baryons: $0 \rightarrow 0 + 0 + 0$ ✓

Leptons: $+1 \rightarrow +1 + (-1) + 1$ ✓

This reaction is possible.

b $n \rightarrow p + e^- + U$

Charge: $0 \rightarrow 1 + -1 + ?$

Baryons: $+1 \rightarrow +1 + 0 + ?$

Leptons: $0 \rightarrow 0 + 1 + ?$

So particle U has no charge, is not a baryon, and has a lepton number of -1.

It is an uncharged antilepton, so an antineutrino, and, as the particle it appears with is first generation, it is an electron antineutrino $\overline{\nu}_e$.

β^- decay

β^- decay was one of three naturally occurring radiations discovered at the end of the 19th century. Deflection experiments confirmed that the β^- particles were in fact fast-moving electrons. (This confirmation was possible only after allowance had been made for the increase of mass of the particle at high speed, as predicted by relativity.)

This created a problem for the new Rutherford model of the atom, since the high energy of the β^- particles could not be explained by transitions between energy levels outside the nucleus. The β^- particles must originate from within the nucleus, but there were known to be only protons and neutrons in the nucleus, no electrons.

The answer is that a neutron within the nucleus changes to a proton. The proton remains within the nucleus and an electron is emitted from the nucleus as a β^- particle:

$$_Z^A X \rightarrow _{Z+1}^A Y + _{-1}^0 e$$

While this satisfied the conservation of mass and charge, it did not give conservation of energy. The β^- particles from a particular isotope were emitted with a continuous range of energies. In 1933 Pauli proposed that a second particle is emitted in the process. This new particle had to have zero charge and zero rest mass but could take off some kinetic energy. This 'neutrino' or 'little neutral one' would by its nature be very difficult to detect, and was first confirmed in the 1950s. The full reaction equation for the above decay is:

$$_Z^A X \rightarrow _{Z+1}^A Y + _{-1}^0 e + _0^0 \overline{\nu}_e$$

Now we can have a further insight into this reaction, and consider what is happening in terms of the fundamental particles within the nucleons. In the β^- decay the neutron (udd) changes to a proton (uud), so in effect a down quark has changed to an up quark. The weak nuclear force induces this change, with the W^- boson exchange particle emitted (see Table 4), subsequently decaying to an electron and an antineutrino:

$$_{-1/3}^{1/3} d \rightarrow _{+2/3}^{1/3} u + W^- \text{ followed by } W^- \rightarrow _{-1}^0 e + _0^0 \overline{\nu}_e$$

This can be illustrated using a **Feynman diagram** (Figure 37).

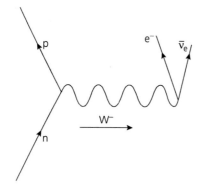

Figure 37 Representation of β⁻ decay

Knowledge check 16

Beta-plus (β⁺) decay can be represented by the equation:

$$p \rightarrow n + e^+ + v_e$$

a Identify the four particles in the reaction, stating whether they are baryons or leptons.

b Explain, in terms of quarks, what has occurred in the reaction.

C Confirm that the reaction equation above conforms with the conservation rules.

Summary

- A fundamental particle is one that has no internal structure.
- There are four fundamental forces which affect the ways that particles interact: gravitational, electromagnetic, strong nuclear and weak nuclear.
- The forces are produced by the exchange of particles known as gauge bosons.
- All observable matter is composed of quarks and leptons, and antimatter is composed of antiquarks and antileptons.
- Particles that are made up of quarks are called hadrons. The quarks combine in twos or threes only, forming mesons and baryons respectively.
- In all particle reactions there must be conservation of charge, baryon number and lepton number.
- β⁻ decay is an example of a reaction in which a lepton pair is created when a quark changes type.

Questions & Answers

The unit assessment

Unit A22 is a written examination of duration 2 hours. It consists of a number of compulsory, short, structured questions. Some of the questions may require an extended response of several sentences. The examination is designed to assess your understanding of all elements in the specification for this unit and all questions must be attempted. It is therefore essential that you revise all sections of the unit. In addition the questions may have elements of synoptic assessment, drawing together different strands of the specification. The questions may not follow the content order of the specification.

The exam incorporates an assessment of the **quality of written communication.** You must ensure your responses are legible and that spelling, punctuation and grammar are accurate. Use well-structured sentences starting with a capital letter and ending with a full stop. Present information clearly, in a logical sequence, and use appropriate scientific language.

Some examination questions will require you to demonstrate your knowledge and understanding of physics, and some questions will require you to apply this understanding to unfamiliar situations. It is important to remember that when presented with an unfamiliar situation, the principles of physics are the same and you have all the tools at your disposal to solve the problem.

Command terms

Examiners use certain words to indicate the type of response expected. (The depth of response required is usually indicated by the number of marks allocated.) It is helpful to be familiar with these terms.

- **Define** — a formal statement. If an equation is given each term used must be identified.
- **State** — a concise answer with little or no supporting argument.
- **Explain** — usually a definition together with some relevant comment on the significance of the term.
- **List** — a number of points or terms with no explanation.
- **Describe** — the key points relating to a particular concept or experiment.
- **Calculate** — a numerical answer is required. Working out should be shown.
- **Measure** — indicates that a quantity can be obtained directly by use of a suitable piece of apparatus.
- **Determine** — indicates that a quantity cannot be obtained directly but rather by calculation, with given or measured values substituted in a suitable form of a known equation.

- **Show** — a calculation has to be performed to obtain an already given value. All work must be shown clearly and logically, with the final value shown to a greater significance than required to confirm that it has not been a 'fudge'.
- **Sketch** — when related to graphs, you are expected to show the general shape or trend of the graph, including intercepts if appropriate. The detail on the axes — quantities, units, origin — is normally expected. Additional values will be specifically asked for if required.
- **Draw** — a carefully and fully labelled diagram, with apparatus shown, using standard symbols, to enable all necessary measurements to be taken. Use a ruler where appropriate and do not make it too small. Note the difference between a circuit diagram and a drawing of the apparatus required.
- **Estimate** – a calculation involving a reasonable assumption of one of the quantities used, leading to an answer of a certain order of magnitude.

Remember the following points:
- State definitions accurately.
- Always write down the formula you are using in a calculation.
- Show all substitutions and working out.
- Check you have included the correct units.
- Use the correct number of significant figures.
- Use a ruler and pencil to draw simple diagrams accurately and neatly.
- Label diagrams fully.
- Know how to use your calculator — you will need it.

Mathematics useful in Unit A2 2

- Standard form is a convenient way of writing down very large or very small numbers. The number is written as a number between 1 and 10 multiplied by 10 to the appropriate power — e.g. quoted to 3 significant figures, $345\,000 = 3.45 \times 10^5$.
- Be familiar with the prefixes that represent decimal multiples and submultiples of units.

Multiplying factor	Prefix	Symbol	Multiplying factor	Prefix	Symbol
10^{-2}	centi	c	10^3	kilo	k
10^{-3}	milli	m	10^6	mega	M
10^{-6}	micro	μ	10^9	giga	G
10^{-9}	nano	n	10^{12}	tera	T
10^{-12}	pico	p			
10^{-15}	femto	f			

- Round numbers to the appropriate number of significant figures:
  - All non-zero digits are significant.
  - Zeros between non-zero digits are significant, 5.06 (3 sig. fig.).
  - Leading zeros are never significant, e.g. 0.0506 (3 sig. fig.).
  - In a number with a decimal point, trailing zeros, those to the right of the last non-zero digit, are significant, e.g. 5.060 (4 sig. fig.).

- In a number without a decimal point, trailing zeros may or may not be significant. More information through additional graphical symbols or explicit information on errors is needed to clarify the significance of trailing zeros, e.g. 74 500 (may be 3, 4 or 5 sig. fig.).
- It is best to quote a number in standard form, i.e. a number between 1 and 10 multiplied by 10 to a power, to best see the number of significant figures.

■ The angle between two lines can be measured in degrees or radians. In a full circle (360°) there are 2π radians. Angles can be converted between degrees and radians as follows:

angle in radians $= \dfrac{2\pi}{360} \times$ angle in degrees

■ Sine and cosine graphs in terms of degrees and radians:

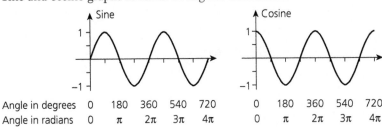

	Sine					Cosine			
Angle in degrees	0	180	360	540	720	0	180	360	540	720
Angle in radians	0	π	2π	3π	4π	0	π	2π	3π	4π

Data

There will be a *Data and Formulae Sheet* inside the Unit A2 2 examination paper. A full list of useful formulae is given on pp. 47–48 of the CCEA specification.

About this section

This section consists of three self-assessment tests. Try the questions without looking at the answers, allowing 2 hours for each test. Then check your responses against the answers and the examiner comments to find out how you might improve upon your performance.

For question parts worth multiple marks, ticks (✔) are included in the answers to indicate where the examiner has awarded marks.

Comments on some questions are preceded by the icon ⊜. They offer tips on what you need to do in order to gain full marks. Some student responses are followed by comments, indicated by the icon ⊜, which highlight where credit is due or could be missed.

■ Self-assessment test 1

Question 1

Physicists use the concept of a field to help explain certain behaviour.

(a) (i) What do you understand by the term **field**? (1 mark)

(ii) Complete Table 1, which compares gravitational and electric fields. (3 marks)

Table 1

	Gravitational field	Electric field
Physical quantity responsible		Charge
Force	Attractive	
Field strength unit	$N\,kg^{-1}$	

(b) (i) Figure 1 represents a positive point charge. Sketch the electric field pattern around the charge. (2 marks)

Figure 1

(ii) Copy the axes of Figure 2 and sketch the graph of field strength magnitude against distance from the point charge. (2 marks)

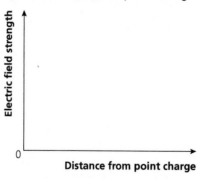

Figure 2

(c) (i) Figure 3 shows parallel plates across which a constant potential difference is maintained. Sketch the electric field pattern between the plates. (2 marks)

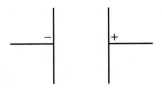

Figure 3

(ii) Copy the axes of Figure 4 and sketch the graph of field strength magnitude against distance from the negative plate.

(1 mark)

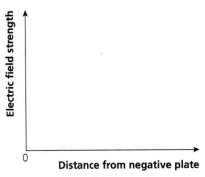

Figure 4

Answer

(a) (i) A region of space where force is experienced ✓

ℯ If more marks had been assigned, a reference to charge or mass or magnetic object may have been expected.

(ii)

	Gravitational field	Electric field
Physical quantity responsible	Mass ✓	Charge
Force	Attractive	Attractive or repulsive ✓
Field strength unit	$N\,kg^{-1}$	$N\,C^{-1}$ ✓

(b) (i) Symmetrical ✓, arrow direction outwards ✓

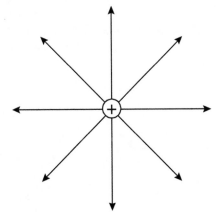

(ii) Correct curve ✓, asymptotic ✓

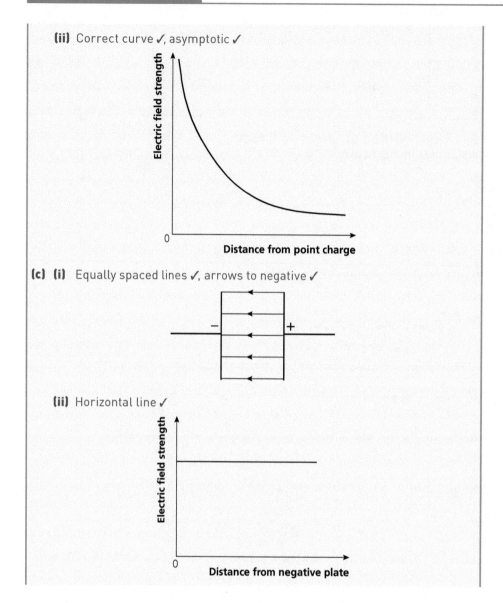

(c) (i) Equally spaced lines ✓, arrows to negative ✓

(ii) Horizontal line ✓

Question 2

A capacitor of capacitance 4.7 µF, a resistor of resistance 33 kΩ and a battery of e.m.f. 6.0 V and negligible internal resistance are connected in series, as shown in Figure 5.

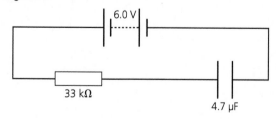

Figure 5

a) Calculate the potential difference across the resistor for the instant at which the charging current in the circuit is 0.07 mA. (3 marks)

b) Deduce the potential difference across the capacitor at this instant. (2 marks)

c) Calculate the charge on the capacitor at this instant. (2 marks)

d) Assuming the current remains approximately constant at 0.07 mA for the next 10 milliseconds, what is the extra charge which flows to the capacitor in this time? (2 marks)

e) What is the new potential difference across the capacitor after the 10 milliseconds? (2 marks)

Answer

(a) $V = IR$ ✓ $= (0.07 \times 10^{-3}) \times (33 \times 10^{3})$ ✓ $= 2.3\,V$ ✓

(b) $V_S = V_R + V_C$ ✓

$\qquad 6.0 = 2.3 + V_C$

$\qquad V_C = 3.7\,V$ ✓

(c) $Q = CV$ ✓ $= (4.7 \times 10^{-6}) \times 3.7 = 17.4 \times 10^{-6}\,C$ ✓

(d) $Q = It$ ✓ $= (0.07 \times 10^{-3}) \times (10 \times 10^{-3}) = 0.7 \times 10^{-6}\,C$ ✓

(e) Total charge has become $18.1 \times 10^{-6}\,C$ ✓

$\qquad$ New potential difference $= \dfrac{(18.1 \times 10^{-6})}{(4.7 \times 10^{-6})} = 3.9\,V$ ✓

Question 3

a) An electron travelling at $6.0 \times 10^{6}\,m\,s^{-1}$ in a vacuum enters a region of uniform magnetic field of flux density 40 mT, as shown in Figure 6.

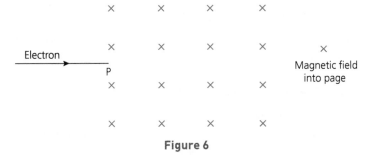

Figure 6

(i) Mark the direction of the force on the electron when it enters the magnetic field at P. (2 marks)

(ii) Calculate the magnitude of the force on the electron. (3 marks)

(iii) Explain why, when the electron is moving in the magnetic field, it follows part of a circular path. (3 marks)

(iv) Calculate the radius of the circle of which this arc is a part. (3 marks)

(b) An ion has mass 6.64×10^{-26} kg and charge 3.20×10^{-19} C. When in a certain uniform electric field it experiences an acceleration of 1.25×10^{11} m s^{-2}, due to the field.

(i) Calculate the magnitude of the field strength of the electric field. (4 marks)

(ii) Determine the potential difference that must be applied across a pair of parallel plates, 5.0 cm apart, to produce this field. (3 marks)

Answer

(a) (i) An arrow vertically ✓ downward at P ✓

(ii) $F = Bqv$ ✓ $= (40 \times 10^{-3}) \times (1.6 \times 10^{-19}) \times (6.0 \times 10^6)$ ✓ $= 3.84 \times 10^{-14}$ N ✓

(iii) Force direction not fixed *or* force direction changes ✓. Force direction always perpendicular to path ✓. Acts as centripetal force ✓.

(iv) $F_C = F_B$ therefore:

$$mv^2/r = Bqv$$ ✓

$$r = \frac{mv}{Bq} = \frac{(9.1 \times 10^{-31}) \times (6.0 \times 10^6)}{(40 \times 10^{-3}) \times (1.6 \times 10^{-19})}$$ ✓

$$r = 8.5 \times 10^{-4} \text{ m}$$ ✓

(b) (i) $F = ma$ ✓ $= (6.64 \times 10^{-26}) \times (1.25 \times 10^{11}) = 8.3 \times 10^{-15}$ N ✓

$$E = \frac{F}{q} ✓ = \frac{(8.3 \times 10^{-15})}{(1.6 \times 10^{-19})} = 5.2 \times 10^4 \text{ N C}^{-1}$$ ✓

(ii) $E = V/d$ ✓

$$V = (5.2 \times 10^4) \times (5.0 \times 10^{-2})$$ ✓ $= 2.6 \times 10^3$ V ✓

Question 4

Figure 7 shows part of a system used to transmit electrical energy over long distances.

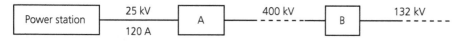

Figure 7

(a) Name the components A and B in Figure 7, and state their functions. (3 marks)

(b) List three possible sources of energy loss in component A, making it clear as to where within the component the loss occurs. (3 marks)

(c) (i) State and explain, using relevant equations, one advantage of transmitting electrical energy at 400 kV rather than the normal domestic supply voltage value of 230 V. (4 marks)

(ii) What is the disadvantage of this higher voltage transmission and how is it resolved? (2 marks)

(d) Assume that component A is 100% efficient. Making use of the information given in Figure 7, calculate, for the 400 kV section of the system, the power loss per metre of length of cable, given that the resistance of 1 metre of the cable is 0.02Ω. (5 marks)

Answer

(a) Both A and B are transformers ✓.

Function of A is to step up voltage ✓.

Function of B is to step down voltage ✓.

(b) Any three of:

Hysteresis in the core

Eddy currents in the core

Resistive heating in the windings

Flux density loss in the core or between windings ✓✓✓

ⓔ Both source and location of energy loss are needed.

(c) (i) To reduce power loss ✓

Explanation: power $P = VI$ ✓, increase in V will decrease I for same P ✓, hence I^2R decreases ✓

(ii) Danger of electrocution ✓
Isolation by use of tall pylons ✓

(d) Since transformer A is 100% efficient, current I in 400 kV cables is given by:

$$(400 \times 10^3) \times I = (25 \times 10^3) \times 120 \checkmark$$

$$I = 7.5\,\text{A} \checkmark$$

Power loss per metre = I^2 × resistance per metre

Power loss per metre = $(7.5)^2 \times 0.02$ ✓ = 1.1 ✓ $\text{W\,m}^{-1}$ ✓

Question 5

(a) What are the SI base units of the universal gravitational constant G? (3 marks)

ⓔ This question would not be asked in this form in an actual exam because the SI unit of G is given with its value on the data sheet.

(b) What are the SI base units of permittivity, ε? (4 marks)

(c) The unit of magnetic flux Φ is the weber; what are its SI base units? (4 marks)

Answer

(a) $F = Gm_1m_2/r^2$, $G = Fr^2/m_1m_2$ ✓

Units of G = $(\text{N})(\text{m}^2)(\text{kg})^{-2}$ ✓

Base units of G = $(\text{kg} \times \text{m} \times \text{s}^{-2})(\text{m}^2)(\text{kg})^{-2} = \text{kg}^{-1}\,\text{m}^3\,\text{s}^{-2}$ ✓

(b) $F = q_1q_2/4\pi\varepsilon r^2$, $\varepsilon = q_1q_2/4\pi Fr^2$ ✓

Units of $\varepsilon = (C^2)(N)^{-1}(m)^{-2}$ ✓

Base units of $\varepsilon = (C)^2(kg \times m \times s^{-2})^{-1}(m)^{-2} = C^2 kg^{-1} m^{-3} s^2$ ✓ $= A^2 kg^{-1} m^{-3} s^4$ ✓

ⓔ It is a common mistake to consider coulomb C rather than ampere A as a base unit.

(c) $\Phi = BA$ ✓, $F = BIL$ ✓, so $\Phi = FA/IL$

Unit of Φ, $Wb = (N)(m^2)/(A)(m)$ ✓

Base units of $\Phi = (kg \times m \times s^{-2})(m^2)(A^{-1})(m^{-1}) = kg\,m^2 s^{-2} A^{-1}$ ✓

Question 6

(a) State the laws of electromagnetic induction. Comment on the connection between one of these laws and the principle of conservation of energy. (3 marks)

(b) A square, flat coil of wire rests on a horizontal table. A uniform magnetic field is directed perpendicularly to the plane of the table, thereby passing through this coil. The flux density of the field can be varied.

(i) The flux density B is changed with time in three different ways, as shown in Figure 8. For each change, sketch a graph on the axes below to show the corresponding variation with time of the induced e.m.f. E in the coil. (6 marks)

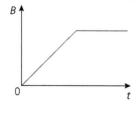 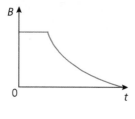 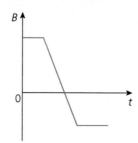

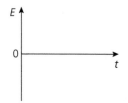 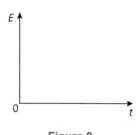 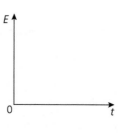

Figure 8

(ii) The coil has 20 turns, and the length of a side is 15 cm. One metre of the wire of the coil has resistance $0.35\,\Omega$. The ends of the coil are connected to a resistor of resistance $5.0\,\Omega$. The magnetic flux density is reduced from $0.06\,T$ to zero in $0.75\,s$. Calculate the current in the resistor. (8 marks)

Answer

(a) Faraday's law: induced e.m.f. is equal to the rate of change of flux linkage ✓.

ℯ An equation would be acceptable if the symbols are identified.

Lenz's law: the direction of the induced e.m.f. or current is such that it opposes the flux change producing it ✓.

Comment: Lenz's law implies that it is not possible to obtain unlimited energy from electromagnetic induction ✓.

(b) (i) First graph: horizontal line ✓ then vertical line to zero ✓
Second graph: zero line ✓ then vertical followed by diagonal line ✓
Third graph: zero line at beginning and end ✓ and non-zero horizontal line between ✓

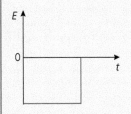 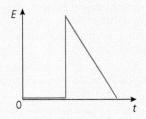 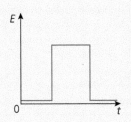

ℯ An additional mark could have been awarded for the directional, +/− aspect ✓.

(ii) Coil resistance = $20 \times 4 \times 0.15 \times 0.35 = 4.2\,\Omega$ ✓
Total circuit resistance = $5.0 + 4.2 = 9.2\,\Omega$ ✓
$\Delta\Phi = \Delta BA$ ✓ $= (0.06) \times (0.15)^2$ ✓

$$E = (-)N\frac{\Delta\Phi}{\Delta t} = \frac{20 \times (0.06) \times (0.15)^2}{0.75}$$ ✓

$E = 3.6 \times 10^{-2}\,\text{V}$ ✓

$$I = \frac{E}{R} \checkmark = \frac{(3.6 \times 10^{-2})}{9.2} = 3.9 \times 10^{-3}$$ ✓

Question 7

(a) (i) Describe the principle of operation of a synchrotron. Mention any relevant design features. (6 marks)

   (ii) Show that, in a synchrotron, the path followed by a particle of mass m and charge q at a speed v has a radius r given by $r = mv/Bq$. (1 mark)

(b) Two of the accelerators built at CERN were the Super Proton Synchrotron (1976–present) and the Large Electron Positron collider (1989–2000).

The Super Proton Synchrotron (SPS) accelerates protons and antiprotons to kinetic energies of 450 GeV.

The Large Electron Positron collider (LEP) accelerated electrons and positrons to kinetic energies of 50 GeV and had a radius of 4.25 km.

(i) Both these accelerators involve particles and their corresponding antiparticles. What is an antiparticle? (2 marks)

(ii) Describe what happens in the LEP when an electron collides with a positron. (3 marks)

(iii) Assuming the same maximum B-field is used in both accelerators, explain, with reference to the equation given in (a)(ii), why the velocity of the particles in the LEP collider is greater than in the SPS accelerator. (2 marks)

Answer

(a) (i) Circular accelerator tube ✓

Charged particles injected and accelerated by electric field ✓

Deflected by magnetic field of a series of (electro)magnets into circular path ✓

Magnetic field strength increased to maintain same orbital radius as their speed increases ✓

Accelerating field provided by alternating p.d. applied across series of cavities ✓

Frequency of p.d. increased to ensure synchronised with particles ✓

(ii) $F_B = F_C$, so $Bqv = mv^2/r$ ✓ and hence $r = mv/Bq$

(b) (i) An identical particle ✓ but with opposite charge, (lepton number and baryon number) ✓

(ii) Annihilation ✓ with the production of two ✓ gamma photons ✓

(iii) $v = Bqr/m$

q: the same for both

r: larger in LEP ✓

m: smaller in LEP ✓

Therefore the velocity of the particles is greater in the LEP (even though the kinetic energy is 9 times smaller).

■ Self-assessment test 2

Question 1

(a) The gravitational field strength on the surface of the Moon is 1.62 N kg^{-1}. The radius of the Moon is 1740 km.

Show that the mass of the Moon is 7.35×10^{22} kg. (2 marks)

(b) The Moon rotates about its axis (while moving in orbit around the Earth). Scientists may wish to place a satellite into a geostationary orbit above the Moon's surface.

(i) What is meant by a geostationary orbit? (1 mark)

(ii) State three features of the motion of a satellite in a geostationary orbit. (3 marks)

(iii) The period of rotation of the Moon about its axis is 27.3 days. Calculate the radius of the geostationary orbit. (4 marks)

c) The times for the Earth and Jupiter to orbit the Sun are 365 days and 4330 days respectively. The radius of the orbit of the Earth around the Sun is 1.50×10^8 km. Calculate the radius of the orbit of Jupiter. (4 marks)

Answer

(a) $g = \dfrac{Gm}{r^2}$ ✓

$M_M = \dfrac{g_M r_M^2}{G} = \dfrac{(1.62) \times (1.740 \times 10^6)^2}{(6.67 \times 10^{-11})}$ ✓ $= 7.35 \times 10^{22}$ kg

(b) (i) Satellite stays in the same position relative to the moon/planet's surface ✓.

(ii) Period equal to the period of rotation of the planet ✓
Orbit in same direction as the direction of rotation of the planet ✓
Orbit in an equatorial plane ✓

(iii) $F_G = F_C$, therefore:

$$\dfrac{Gm_s m_M}{r^2} = m_s r\omega^2 ✓$$

$$r^3 = \dfrac{Gm_M}{\omega^2}$$

Using $T = 2\pi/\omega$ ✓:

$$r^3 = \dfrac{Gm_M T^2}{4\pi^2} = \dfrac{(6.67 \times 10^{-11}) \times (7.35 \times 10^{22}) \times (27.3 \times 24 \times 60 \times 60)^2}{4\pi^2} ✓$$

$r = 8.84 \times 10^7$ m or 8.84×10^4 km ✓

(c) $T^2/r^3 = $ constant ✓

$T_E^2/r_E^3 = T_J^2/r_J^3$ ✓

$r_J^3 = (4330)^2 \times (1.50 \times 10^8)^3/(365)^2$ ✓

$r_J = 7.80 \times 10^8$ km ✓

ℯ You have not been directed to answer in a specific unit, so you can work in km or m, so long as you are consistent and clearly label your answer.

Question 2

(a) (i) Define electric field strength. (1 mark)

(ii) Electric field strength is a vector quantity. Explain how its direction is defined. (2 marks)

(b) (i) According to the Bohr model of the atom, electrons can revolve around the nucleus only in certain allowed orbits. For hydrogen, an electron in its ground state will orbit the nucleus with a radius of 5.29×10^{-13} m. Calculate the magnitude of the electric force experienced by this electron. (4 marks)

(ii) State whether the force is attractive or repulsive and explain your answer. (2 marks)

(c) (i) Charges of equal magnitude and sign, as shown in Figure 1, are placed at two of the vertices of the equilateral triangle ABC. Show the direction of the electric field at B. (2 marks)

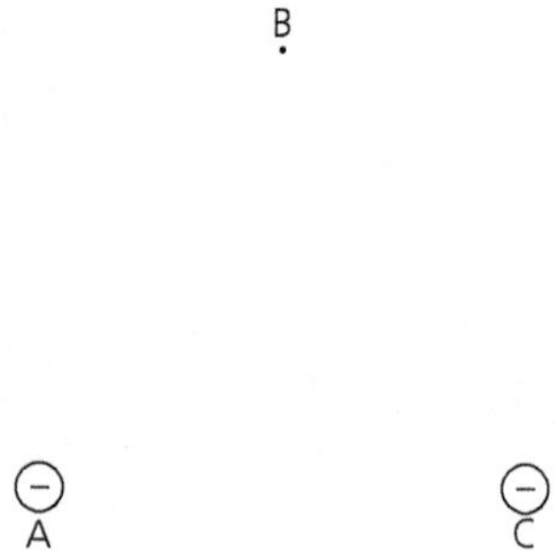

B
.

⊖
A

⊖
C

Figure 1

(ii) Charges of equal magnitude and opposite sign, as shown in Figure 2, are placed at two of the vertices of the equilateral triangle RST. Show the direction of the electric field at R. (2 marks)

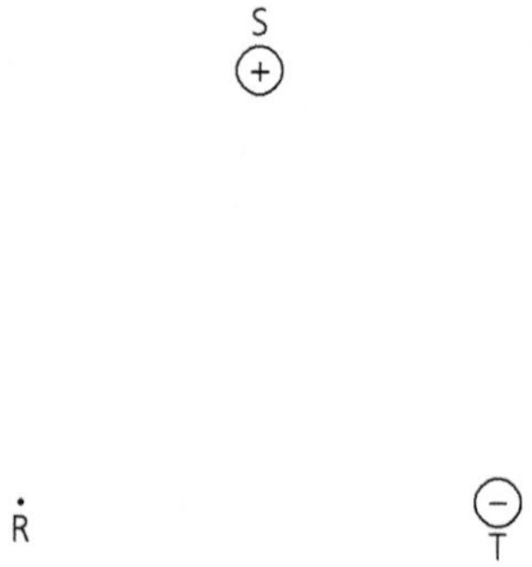

S
⊕

.
R

⊖
T

Figure 2

(iii) A point charge at M causes a certain electric field strength E at X, 4.0 m from M (Figure 3). If the point charge at M is tripled, an equal field strength E is created further away, at Y. Calculate the distance XY. (4 marks)

Figure 3

Answer

(a) (i) The force per coulomb ✓

(ii) The direction in which a small positive charge ✓ would move if placed in the field ✓

(b) (i) There is one proton in the nucleus of hydrogen ✓. So:

$$F = \frac{kq_e q_p}{r^2} \checkmark = \frac{(8.99 \times 10^9) \times (1.60 \times 10^{-19}) \times (1.60 \times 10^{-19})}{(5.29 \times 10^{-13})^2} \checkmark$$

$$F = 8.22 \times 10^{-4} \text{ N} \checkmark$$

ⓔ Don't mix up the constant k with the Boltzmann constant.

(ii) Attractive ✓, unlike charges attract ✓

(c) (i) Two individual forces with correct direction ✓, resultant with correct direction ✓

(ii) Two individual forces with correct direction ✓, resultant with correct direction ✓

ⓔ Usually the length of the resultant should indicate the relative magnitude of the force, but here only the direction is asked for.

(iii) $E = \dfrac{kq}{r^2}$ ✓

Let the distance MY = d

$$E = \frac{kq}{4.0^2} = \frac{k(3q)}{d^2} \checkmark$$

$$d^2 = 3 \times (4.0)^2 = 48, \ d = 6.9 \text{ m} \checkmark$$

$$XY = 6.9 - 4.0 = 2.9 \text{ m} \checkmark$$

Question 3

(a) (i) State what is meant by a positive ion.　　　　　　　　　　　　　　(1 mark)

(ii) State what is meant by the term isotope.　　　　　　　　　　　　　(1 mark)

(b)

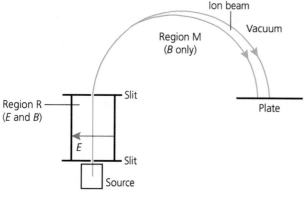

Figure 4

(i) Figure 4 shows a fine beam of positive ions with a particular speed travelling undeflected in a region R where an electric field of strength $2.6 \times 10^4\,\text{V}\,\text{m}^{-1}$ *and* a magnetic field of flux density 0.50 T act at right angles to each other and to the direction of the beam. Calculate the speed of the ions if each carries a single unit of charge.　　　(3 marks)

(ii) The ion beam enters the region M where only the magnetic field acts, and the beam is deflected in a semicircular path as shown. Find an expression for the radius r of the path involving the mass m of the ions in the beam.　　　(2 marks)

(iii) The direction of the B-field is the same across the whole apparatus. Determine the direction of this B-field, explaining how you have come to this conclusion.　　　(3 marks)

(iv) A photographic plate is placed in the path of the beam in the position shown. When the plate is processed, it shows two lines separated by a small distance. If the ion beam contains two isotopes of mass $4.0 \times 10^{-26}\,\text{kg}$ and $4.4 \times 10^{-26}\,\text{kg}$ respectively, calculate the separation of the lines on the plate.　　　(5 marks)

Answer

(a) (i) An atom that has lost one or more electrons ✓

(ii) Any of two or more forms of a chemical element, having the same number of protons in the nucleus, or the same atomic number, but having different numbers of neutrons in the nucleus ✓

(b) (i) For no deflection $F_E = F_B$, so $qE = Bqv$ ✓
Therefore:

$$v = \frac{E}{B}\,✓ = \frac{(2.6 \times 10^4)}{0.5} = 5.2 \times 10^4\,\text{m}\,\text{s}^{-1}\,✓$$

(ii) F_B = centripetal force, so:

$$Bqv = \frac{mv^2}{r} \checkmark \text{ and } r = \frac{mv}{Bq} \checkmark$$

(iii) Positive ion beam so current direction initially vertically upwards ✓
Magnetic force direction opposite to E-field in region R, i.e. to the right, or
magnetic force must be to the right to cause the circular path shown ✓
Using Fleming's left-hand rule, B-field is out of the plane of the page ✓

(iv) Using $r = mv/Bq$:

When $m = 4.0 \times 10^{-26}$ kg:

$$r = \frac{(4.0 \times 10^{-26}) \times (5.2 \times 10^4)}{0.50 \times (1.6 \times 10^{-19})} \checkmark$$

$r = 2.60 \times 10^{-2}$ m ✓

When $m = 4.4 \times 10^{-26}$ kg:

$$r = \frac{(4.4 \times 10^{-26}) \times (5.2 \times 10^4)}{0.50 \times (1.6 \times 10^{-19})}$$

(or by ratios since $r \propto m$)

$r = 2.86 \times 10^{-2}$ m ✓

The separation equals the difference in diameters ✓ = 0.52 cm ✓

Question 4

(a) Two capacitors of capacitance 2.0 μF and 200 pF are each given a charge of 4.0 nC.
Calculate:

   (i) the potential difference across each capacitor   (3 marks)

   (ii) the energy stored in each capacitor   (3 marks)

(b) Three capacitors with capacitances of 0.22 μF, 0.47 μF and 1.0 μF are available.

   (i) What is the largest value of capacitance which can be produced using
these three capacitors?
Draw a circuit for the combination which achieves this.   (2 marks)

   (ii) What is the smallest value of capacitance which can be produced using
these three capacitors?
Draw a circuit for the combination which achieves this.   (3 marks)

(c) A capacitor C of capacitance 2 μF may be connected either to a 100 V d.c.
supply, or to a resistor R of resistance 1 MΩ, by means of a two-way switch S.
Figure 5 shows the capacitor connected to the d.c. supply.

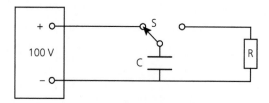

Figure 5

After a short time the switch is changed over to connect the capacitor to the resistor R.

(i) State the name of the quantity given by the product of the capacitance of the capacitor and the resistance of the resistor, CR, in such a circuit, state its physical significance and determine its value in this case. **(3 marks)**

(ii) A voltmeter is used to measure the potential difference across the capacitor. Determine the values indicated by the voltmeter at 2 s intervals for the first 10 seconds from the time the switch is changed over. Draw up a table to show these values. **(3 marks)**

(iii) Making use of the table of values in (c)(ii), draw a graph of the potential difference across the capacitor against time. From your graph determine the time taken from the changeover of the switch for the potential difference to fall to 20 V. **(4 marks)**

(iv) Confirm, or otherwise, the time taken for the potential difference to fall to 20 V by using the equation $V = V_0 e^{-t/CR}$. **(2 marks)**

Answer

(a) (i) $Q = CV$, so:

$$V = \frac{Q}{C} \checkmark$$

For the 2.0 µF capacitor:

$$V = \frac{(4.0 \times 10^{-9})}{2(2.0 \times 10^{-6})} = 2.0 \times 10^{-3} = 2.0\,\text{mV} \checkmark$$

For the 200 pF capacitor:

$$V = \frac{(4.0 \times 10^{-9})}{2(200 \times 10^{-12})} = 20\,\text{V} \checkmark$$

(ii) $E = \frac{Q^2}{2C} \checkmark$

ⓔ It is best to use this form of the equation as then you are not relying on values you have calculated.

For the 2.0 µF capacitor:

$$E = \frac{(4.0 \times 10^{-9})^2}{2(2.0 \times 10^{-6})} = 4.0 \times 10^{-12} \qquad \text{J} \checkmark$$

For the 200 pF capacitor:

$$E = \frac{(4.0 \times 10^{-9})^2}{2(200 \times 10^{-12})} = 4.0 \times 10^{-8} \quad J \checkmark$$

(b) (i) $C = 0.22 + 0.47 + 1.00 = 1.69\,\mu F$ ✓

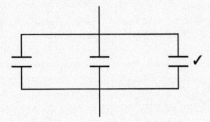

(ii) $\dfrac{1}{C} = \dfrac{1}{0.22} + \dfrac{1}{0.47} + \dfrac{1}{1.00} = 7.67$ ✓

$C = 0.13\,\mu F$ ✓

(c) (i) Time constant ✓

This is the time taken for the potential difference across the capacitor (or its charge or the current) to fall to 37% of the initial value on discharging the capacitor ✓

$CR = (2 \times 10^{-6}) \times (1 \times 10^{6}) = 2\,s$ ✓

(ii) Each interval of 2 s is one time constant, so p.d. falls by 37% each time.

Time/s	0	2	4	6	8	10
p.d./V	100	37.0	13.7	5.1	1.9	0.69

First value of 100 V ✓, correct use of 37% on one occasion ✓, six values in agreement ✓

(iii)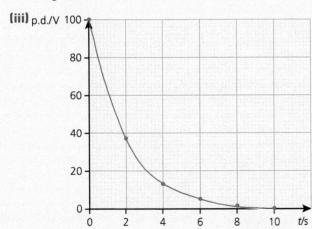

Correct scales ✓, correct points ✓, smooth curve ✓, time to fall to 20 V, 3.1–3.3 s ✓

(iv) $V = V_0 e^{-t/CR}$

$\dfrac{V_0}{V} = e^{t/CR}$ and $\ln\left(\dfrac{V_0}{V}\right) = \dfrac{t}{CR}$ ✓

When $V = 20\,\text{V}$, $V_0/V = 5$

$\ln 5 = t/2$ and so $t = 3.2\,\text{s}$ ✓

Question 5

(a) A certain a.c. generator has a coil area of $125\,\text{cm}^2$. The coil has 240 turns. The generator produces a peak voltage of 25.0 V when the coil is turning at 6000 revolutions per minute.

   (i) Calculate the frequency of the output.   (1 mark)

   (ii) Sketch a graph showing how the output voltage V varies with time t. Label the time axis with appropriate values.   (2 marks)

   (iii) Sketch a graph showing how the flux linkage Φ varies in the same time t.   (3 marks)

   (iv) Calculate the magnetic flux density required.   (4 marks)

(b) The rate of rotation of the coil is reduced to 4200 revolutions per minute.

   (i) Calculate the new peak output voltage.   (2 marks)

   (ii) Calculate the new output frequency.   (2 marks)

Answer

(a) (i) $f = 6000/60 = 100\,\text{Hz}$ ✓

   (ii) Sine or cosine curve ✓, period 0.01 s. ✓

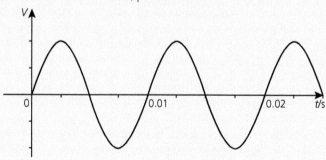

   (iii) Cosine or sine curve ✓, 90° phase difference from V graph in (ii) ✓, same period ✓

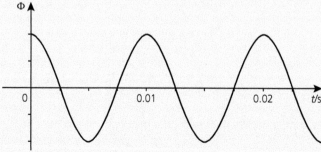

(iv) Maximum voltage $E_0 = BAN\omega$ ✓

$\omega = 2\pi f$ ✓

$$B = \frac{E_0}{AN\omega} = \frac{25.0}{(125 \times 10^{-4}) \times (240) \times (2\pi \times 100)} \checkmark$$

$B = 13.26\,mV$ ✓

(b) (i) $E_0' = 25.0 \times \left(\dfrac{4200}{6000}\right)$ ✓

$E_0' = 17.5\,V$ ✓

(ii) $f' = 100 \times \left(\dfrac{4200}{6000}\right)$ ✓

$f' = 70\,Hz$ ✓

ⓔ Solutions are shown here using ratios, but you could recalculate using the full equations.

Question 6

(a) Consider a neutral atom of sodium ($^{23}_{11}$Na). State how many of each of the following types of particle a neutral atom of sodium contains. Explain your answers.

   (i) leptons   (2 marks)

   (ii) baryons   (2 marks)

   (iii) mesons   (2 marks)

(b) Of the strong and the weak fundamental forces, which are experienced by:

   (i) baryons   (1 mark)

   (ii) leptons   (1 mark)

(c) (i) What is the group name given to particles such as baryons and mesons that are made up of quarks?   (1 mark)

   (ii) State the charge of an up quark and of a down quark.   (2 marks)

   (iii) How many quarks are there in a baryon and in a meson?   (2 marks)

   (iv) When quarks combine together to make particles, what is their possible total relative charge?   (1 mark)

(d) Describe, in terms of quarks and virtual particle exchange, the process of β^- decay.   (3 marks)

Answer

(a) (i) 11 ✓, neutral atom so same number of electrons as protons, and electrons are leptons ✓

   (ii) 23 ✓, the mass number is the number of protons and neutrons, which are the baryons ✓

   (iii) 0 ✓, there are no mesons in the atom ✓

(b) (i) Baryons experience both the strong and the weak nuclear force ✓

(ii) Leptons experience the weak nuclear force ✓

(c) (i) Hadrons ✓

(ii) Up quark $+\frac{2}{3}e$ ✓, down quark $-\frac{1}{3}e$ ✓

(iii) 3 quarks in a baryon ✓, 2 quarks (quark and an antiquark) in a meson ✓

(iv) Possible total relative charges of 0, +1, −1 ✓

(d) Down quark changes to an up quark ✓

W⁻ exchange particle emitted ✓

W⁻ rapidly decays to an electron and (electron) antineutrino ✓

ⓔ Remember that relative charge is its value compared to the elementary charge $e = 1.6 \times 10^{-19}\,C$.

■ Self-assessment test 3

This test provides questions with emphasis on the synoptic and 'stretch and challenge' aspects.

Question 1

(a) (i) State the equation, which includes permittivity, for the electric field strength, E_E, due to a point charge, identifying all the other terms used in the equation. (2 marks)

(ii) What will happen to the force between two charges when a material of high permittivity replaces a vacuum? (1 mark)

(b) Two point charges of +4.8 µC and −5.8 µC are placed at points A and B respectively in a vacuum, 0.15 m apart. See Figure 1.

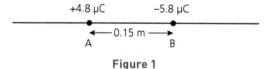

Figure 1

It is required to find a point Z at which the resultant electric field strength due to these two charges is zero.

(i) Explain why Z cannot lie off the line, or its extension, joining the charges. (2 marks)

(ii) Decide whether Z can lie to the left of A, between A and B, to the right of B, or in more than one of these regions. Give reasons for your answer. (5 marks)

(iii) Calculate the location of point Z. (5 marks)

ⓔ Part (b) would be viewed as the 'stretch and challenge' part of the paper.

Answer

(a) (i) $E_E = \dfrac{q}{4\pi\varepsilon r^2}$ ✓

q = value of the charge, r = distance between the point charge and the position in the field where the field strength is being measured, ε = permittivity of medium ✓

(ii) The force will reduce considerably in size (because force $F \propto E_E$ and so $F \propto 1/\varepsilon$) ✓.

(b) (i) If Z was not on line AB, then the forces due to each charge would be in a different and non-parallel direction ✓, and it is then impossible to have a zero resultant ✓.

(ii) Z to the left of A: a positive test charge will be repelled by A ✓ but although B is further away, the value of the charge is greater ✓, so the attractive force could equal the repulsive force. Hence possible ✓.
Z between A and B: a positive test charge will be repelled by A and attracted by B, so field is never zero. Hence impossible ✓.
Z to the right of B: attraction by the larger, nearer charge at B will always dominate, so field is never zero. Hence impossible ✓.

(iii) Let Z be a distance x to the left of A. For zero field at Z:

$$\frac{kq_A}{r_{ZA}^2} = \frac{kq_B}{r_{ZB}^2} \text{ ✓}$$

$$\frac{k(4.8)}{x^2} = \frac{k(5.8)}{(x+0.15)^2} \text{ ✓}$$

$$\frac{(x+0.15)^2}{x^2} = \frac{5.8}{4.8} \text{ ✓}$$

$$\frac{x+0.15}{x} = \left(\frac{5.8}{4.8}\right)^{1/2} \text{ ✓}$$

$$x + 0.15 = 1.1x$$

$$x = 1.5\,\text{m to the left of A ✓}$$

Question 2

(a) A transformer is located at a substation where power is distributed to domestic users. The potential difference is reduced in the substation from 11 kV to 230 V.

(i) What is the ratio of turns on this transformer? (1 mark)

(ii) Which coil has the greater number of turns, primary or secondary? (1 mark)

(b) Electrical power P is to be delivered at an output voltage V to a transmission system. Figure 2 is a schematic diagram of the system.

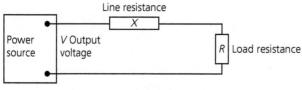

Figure 2

The line resistance, which gives rise to power loss, is X, and the constant load resistance is R.

(i) Write down an expression for the current I in the transmission line in terms of the quantities labelled in Figure 2. (2 marks)

(ii) Show that, for a given power P delivered by the source, the power loss in the lines is inversely proportional to V^2, where V is the output voltage of the source. (3 marks)

(iii) The system of Figure 2 is to transmit power of 200 kW from the source. The total resistance of the transmission lines is 0.80 Ω. It is required that the power loss in the lines should not exceed 0.02% of the total power. Find the minimum output voltage to achieve this. (3 marks)

(c) One of the sources of power loss in a transformer is the generation of heat due to eddy currents in the core of the transformer. A way of reducing eddy currents is to make the core from a number of thin laminations of sheet iron, each electrically insulated from the next by a layer of paint or varnish.

(i) Suggest how this construction reduces eddy current loss. (1 mark)

(ii) State two other sources of power loss associated with the core of the transformer. (2 marks)

Answer

(a) (i) Turns ratio = p.d. ratio = $\dfrac{11000}{230}$ = 47.8 ✓

or the inverse

(ii) Greater number of turns on primary ✓ (a 'step-down' transformer)

(b) (i) $I = \dfrac{V}{R_{tot}}$ ✓ therefore $I = \dfrac{V}{(X + R)}$ ✓

(ii) Power loss = I^2X ✓

Output power from source = P

Using $I = P/V$ ✓:

$$\text{power loss} = \left(\dfrac{P}{V}\right)^2 X = \dfrac{P^2X}{V^2} ✓$$

Loss is inversely proportional to V^2.

(iii) Maximum power loss = 0.02% of 200 kW = 40 W ✓

$$\text{power loss} = \frac{P^2 X}{V^2} = \frac{(200 \times 10^3)^2 \times 0.80}{V^2} = 40 ✓$$

giving minimum output voltage = 28 300 V or 28.3 kV ✓

(c) (i) Increases the resistance of the core, so reducing current ✓

(ii) Hysteresis ✓; flux leakage ✓

Question 3

a) Figure 3 is a graph of the relation between the charge Q on the plates of a capacitor and the potential difference V between them.

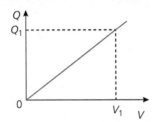

Figure 3

(i) State how the capacitance C of the capacitor may be obtained from Figure 3. (1 mark)

(ii) When the potential difference between the plates of the capacitor is V_1, the corresponding charge is Q_1. What is the equation relating Q_1, V_1 and the energy E_{cap} stored in the capacitor at this potential difference value? State how this energy, E_{cap}, is related to the graph in Figure 3. (3 marks)

(iii) State the equation for the energy E required to move a charge Q_1 through a potential difference V_1. (1 mark)

(iv) Your answers for E_{cap} in (ii) and E in (iii) should be different. Making reference to the process of charging a capacitor, account for this difference. (3 marks)

b) An isolated capacitor A of capacitance 220 μF carries a charge of 5.5 mC. It is then connected to a second, uncharged, capacitor B of capacitance 470 μF, as shown in Figure 4.

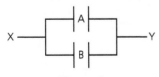

Figure 4

(i) Calculate the original potential difference across the charged capacitor A. (3 marks)

(ii) Calculate the capacitance C of the circuit in Figure 4, measured between points X and Y. (2 marks)

(iii) Calculate the potential difference V between points X and Y. (2 marks)

(iv) Calculate the charge on each capacitor when they are connected as in Figure 4. (2 marks)

> **Answer**
>
> **(a) (i)** Capacitance C = gradient of the line ✓
>
> **(ii)** $E_{cap} = \frac{1}{2}Q_1V_1$ ✓
> E_{cap} = area between the line and the Q-axis ✓ up to (V_1, Q_1) ✓
>
> **(iii)** $E = Q_1V_1$ ✓
>
> **(iv)** When charging a capacitor, charge is continually being added to the plate ✓
> As the charge builds up, the p.d. increases ✓
> So in the calculation of E_{cap}, p.d. is not constant, so the average value $0 \rightarrow V_1$, i.e. $\frac{1}{2}V_1$, is needed ✓
>
> **(b) (i)** $V = \dfrac{Q}{C}$ ✓ $= \dfrac{(5.5 \times 10^{-3})}{(220 \times 10^{-6})}$ ✓ $= 25\,\text{V}$ ✓
>
> **(ii)** $C = C_A + C_B$ when in parallel ✓
> $C = 220 + 470 = 690\,\mu\text{F}$ ✓
>
> **(iii)** $V = \dfrac{Q}{C} = \dfrac{(5.5 \times 10^{-3}}{(690 \times 10^{-4}}$ ✓
>
> $V = 8.0\,\text{V}$ ✓
>
> **(iv)** On capacitor A:
>
> $Q_A = C_A V = (220 \times 10^{-6}) \times 8.0 = 1.76 \times 10^{-3}\,\text{C}$ ✓
>
> On capacitor B:
>
> $Q_B = C_B V = (470 \times 10^{-6}) \times 8.0 = 3.76 \times 10^{-3}\,\text{C}$ ✓

e Note Q_{total} is unchanged, $(1.76 + 3.76) = 5.52\,\text{mC}$. (The additional $0.02\,\text{mC}$ is due to 'rounding up' the V value.)

Question 4

Figure 5 shows two horizontal plates. The region between and around the plates is evacuated.

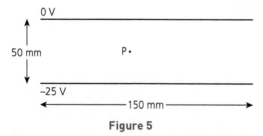

Figure 5

The plates are 150 mm long and are 50 mm apart. The lower plate is maintained at a potential of –25 V with respect to the upper plate.

(a) Calculate the magnitude of the electric field strength at the point P midway between the plates. (2 marks)

(b) What is the direction of the electric field at P? (1 mark)

(c) An electron is situated at point P. Calculate the magnitude of the electric force acting on the electron. (2 marks)

(d) Figure 6 shows the path of an electron beam entering the region between the plates, along a line midway between the plates. Assume the electric field is uniform between the plates and zero elsewhere.

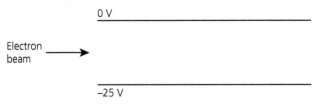

Figure 6

The electrons in the beam have a velocity of $7.5 \times 10^6 \, \text{m s}^{-1}$ as they enter the region between the plates.

(i) Sketch the path of the electron beam as it passes between and beyond the plates. (3 marks)

(ii) Calculate the acceleration of an electron in the beam as it passes between the plates. State the direction of the acceleration. (3 marks)

(iii) Calculate the time taken for an electron in the beam to pass from one end of the plates to the other. (2 marks)

(iv) Use your answers to parts (d)(ii) and (iii) to calculate the distance from the mid-line at which the electron beam leaves the region between the plates. (3 marks)

ⓔ Note the synoptic aspect of this question, requiring the equations of motion and projectile physics from AS.

Answer

(a) $E = \dfrac{V}{d}$ ✓ $= \dfrac{25}{(50 \times 10^{-3})} = 500 \, \text{V m}^{-1}$ ✓

(b) Vertically downwards ✓

(c) $F = QE$ ✓ $= (1.6 \times 10^{-19}) \times 500 = 8.00 \times 10^{-17} \, \text{N}$ ✓

(d) (i) Between the plates, a curved path ✓
Curving upwards ✓
Beyond the plates, a straight line continuation ✓

(ii) $a = \dfrac{F}{m} = \dfrac{(800 \times 10^{-17})}{(9.1 \times 10^{-31})}$ ✓

$a = 8.8 \times 10^{13} \, \text{m s}^{-2}$ ✓

upwards ✓

(iii) Consider the motion in the horizontal plane (velocity remains constant):

$$t = \frac{s_h}{v_h} = \frac{(150 \times 10^{-3})}{(75 \times 10^6)} \checkmark = 2.0 \times 10^{-8}\,s \checkmark$$

(iv) Consider the motion in the vertical plane:

ⓔ Note the time t is the same for the horizontal and the vertical motion.

$$s_v = ut + \tfrac{1}{2}at^2 \checkmark$$

$$s_v = 0 + \tfrac{1}{2}(8.8 \times 10^{13}) \times (2.0 \times 10^{-8})^2 \checkmark$$

$$s_v = 1.76 \times 10^{-2}\,m \text{ or } 18\,mm \checkmark$$

Question 5

(a) A source of protons is located in a vacuum in a region of uniform magnetic field. The protons are emitted from the source with a range of energies, and move at right angles to the direction of the magnetic field, travelling in circular paths.

(i) Show that the frequency f of the circular motion of a proton is given by $f = Be/2\pi m$, where B is the flux density of the magnetic field, e is the elementary charge, and m is the proton mass. (5 marks)

(ii) For a proton with greater energy, how would the values of the following quantities change, if at all? Explain your answers.

Radius of orbit

Frequency of orbit (3 marks)

(iii) The flux density of the uniform magnetic field is 0.40 T. Calculate the frequency f. (2 marks)

(iv) Calculate the kinetic energy with which a proton must be travelling in this magnetic field so that the radius of its path is 240 mm. Give your answer in MeV. (6 marks)

(v) Explain why, in the calculations in (iii) and (iv), it is acceptable to use the rest mass value for the proton. (2 marks)

(vi) Suppose the proton source is replaced by an electron source. Consider an electron travelling with speed v, and compare its path with that of a proton also travelling with speed v, by stating one similarity and two differences between the paths of the two particles. (3 marks)

(b) A narrow beam of identical negatively charged ions enters a uniform magnetic field with constant velocity. The direction of the magnetic field is perpendicular to the path of the ions. As a result, the ions travel in a semicircular path in the region of the magnetic field, and then leave the magnetic field in a direction parallel to that in which the beam entered. The incomplete path of the ions is shown in Figure 7.

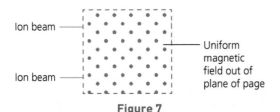

Figure 7

(i) State and explain the direction of travel of the ion beam at the top and at the bottom of Figure 7. (2 marks)

(ii) While the ions are in the magnetic field region, does their momentum change? Explain your answer. (2 marks)

ⓔ This question is longer than usual, with many synoptic elements, involving magnetic field, circular motion, kinetic energy, electron volt, momentum, vectors and relativistic mass.

Answer

(a) (i) $F_B = Bqv$ ✓ with $q = e$

$F_B = F_C$ therefore:

$$Bev = \frac{mv^2}{r} \text{ ✓ and so } v = \frac{Ber}{m} \text{ ✓}$$

Using $v = 2\pi r/T$ and $f = 1/T$ ✓ gives:

$$2\pi rf = \frac{Ber}{m} \text{ ✓ and hence } f = \frac{Be}{2\pi m}$$

(ii) Greater (kinetic) energy, same mass, so speed greater ✓

Radius increased since $v = Ber/m$ ✓

Frequency unchanged since $f = Be/2\pi m$ ✓

(iii) $f = Be/2\pi m$

$$f = \frac{0.40 \times (1.60 \times 10^{-19})}{2\pi \times (1.67 \times 10^{-27})} \text{ ✓}$$

$f = 6.1 \times 10^6$ Hz ✓

(iv) Using $v = Ber/m$ ✓ or $v = 2\pi rf$:

$$v = \frac{0.40 \times (1.60 \times 10^{-19}) \times (240 \times 10^{-3})}{1.67 \times 10^{-27}} \text{ ✓}$$

or

$$v = 2\pi \times (240 \times 10^{-3}) \times (6.1 \times 10^6)$$

$v = 9.2 \times 10^6 \, \text{m s}^{-1}$ ✓

$E_k = \frac{1}{2}mv^2$ ✓

$E_k = 0.5 \times (1.67 \times 10^{-27}) \times (9.2 \times 10^6)^2 = 7.07 \times 10^{-14}$ J ✓

$$7.07 \times 10^{-14} \, \text{J} = \frac{(7.07 \times 10^{-14})}{(1.60 \times 10^{-19}) \times 10^6} = 0.44 \, \text{MeV ✓}$$

(v) The speed is $(9.2 \times 10^6)/(3.0 \times 10^8) = 3\%$ of the speed of light ✓ so relativistic mass increase is negligible ✓.

(vi) Similarity: both follow circular paths ✓.

Differences: opposite direction of rotation ✓; the electron has path of much smaller radius *or* higher frequency of rotation ✓.

(b) (i) In at the bottom, out at the top ✓, by use of Fleming's left-hand rule, with conventional current opposite to negative ion movement ✓.

(ii) Momentum = mass × velocity; velocity is a vector ✓.
Direction is changing, so yes ✓.

Question 6

The energy equivalent of a particle's rest mass is called its 'rest energy'.

(a) Calculate the rest energy of a proton, in GeV. (3 marks)

(b) A particle accelerator is designed to accelerate antiprotons through an effective potential difference of 4.0 GV and make them collide with protons of equal energy moving in the opposite direction. In such a collision, a proton–antiproton pair is created as represented by the equation:

$$p + \bar{p} \rightarrow p + \bar{p} + p + \bar{p}$$

(i) State how the antiproton differs from the proton. (1 mark)

(ii) State the rest energy of the antiproton. (1 mark)

(iii) What is the total kinetic energy of the particles, in GeV, before collision? (2 marks)

(iv) Calculate the total kinetic energy of the particles, in GeV, after the collision. (3 marks)

(v) Explain the advantage of colliding beam experiments rather than high-energy particles hitting a fixed target. (3 marks)

Answer

(a) $E = mc^2$ ✓

$E = (1.67 \times 10^{-27}) \times (3 \times 10^8)^2 = 1.50 \times 10^{-10} \, \text{J}$ ✓

$1 \, \text{eV} = 1.6 \times 10^{-19} \, \text{J}$, therefore:

$E = 1.50 \times 10^{-10} \, \text{J} = 9.40 \times 10^8 \, \text{eV} = 0.94 \, \text{GeV}$ ✓

(b) (i) Proton positively charged, antiproton negatively charged ✓

(ii) Same rest energy as the proton, 0.94 GeV ✓

(iii) $W = QV$, the work done and hence kinetic energy gained is the product of the charge of the particle and the potential difference it has moved through ✓.
Both particles have gained 4.0 GeV
Therefore total kinetic energy = 8.0 GeV ✓

(iv) Energy needed to create proton–antiproton pair = 2 × 0.94 GeV = 1.88 GeV ✓
Kinetic energy remaining = 8.0 – 1.88 = 6.12 GeV ✓

(v) In the colliding beam both of the particle beams have utilised the accelerating capacity of the accelerator, so the available energy is effectively doubled ✓.

Evidence of an explanation of the advantage of higher energy, in terms of momentum:

A fast-moving particle has high momentum, and when it hits a fixed target, the same value of momentum must exist after collision. If the particles have momentum they must have velocity and so kinetic energy. This kinetic energy cannot therefore be available during collision to create new mass ✓.

The total momentum of two colliding particles before collision is zero. So in theory all the energy is available to produce new matter as the total momentum after collision is also zero ✓.

Knowledge check answers

Knowledge check answers

1 $F = GM_E M_M/r^2 = [(6.67 \times 10^{-11}) \times (6.0 \times 10^{24}) \times (7.4 \times 10^{22})]/(3.8 \times 10^8)^2 = 2.1 \times 10^{20}\,N$

2 $g = GM/r^2 = [(6.67 \times 10^{-11}) \times (2.0 \times 10^{30})]/(7.0 \times 10^8)^2 = 2.7 \times 10^2\,m\,s^{-2}$

3 $F = kq_1 q_2/r^2 = ke^2/r^2 = [(9.0 \times 10^9) \times (1.6 \times 10^{-19})^2]/(1.2 \times 10^{-10}) = 1.6 \times 10^{-8}\,N$

4 $V \equiv J\,C^{-1}$ and $J \equiv N\,m$, so $V\,m^{-1} \equiv (N\,m\,C^{-1}) \times m^{-1} \equiv N\,C^{-1}$

5 $C = Q/V = (4.0 \times 10^{-3})/50 = 8.0 \times 10^{-5}\,F = 80\,\mu F$

6 $E = \frac{1}{2}CV^2 = 0.5 \times (470 \times 10^{-6}) \times 12^2 = 3.4 \times 10^{-2}\,J$

7 Series:

$1/C = (1/200) + (1/300) = 5/600$

Therefore $C = 120\,\mu F$

$Q = CV = 120 \times 6 = 720\,\mu J$ or $7.2 \times 10^{-4}\,J$

Parallel:

$C = 200 + 300 = 500\,\mu F$

$Q = CV = 500 \times 6 = 3000\,\mu J$ or $3.0 \times 10^{-3}\,J$

The parallel arrangement will store more charge.

8 $C = Q/V$ and $R = V/I$, so $CR = Q/I$

But $Q = It$, so $CR = Q/I = It/I = t$

9 To lift we need $F = W = mg$. Maximum F is when the wire is placed perpendicular to the horizontal component of the Earth's field.

Then $F = BIL$, therefore $BIL = mg$ and $I = mg/BL$

$I = [(4.5 \times 10^{-3}) \times 9.8]/[(18 \times 10^{-6}) \times 0.1] = 2.5 \times 10^4\,A$

Note that this is a dangerously large current, which would most likely melt the wire.

10 Flux linkage $= NBA = NB\pi r^2 = (25) \times (90 \times 10^{-3}) \times \pi \times (5 \times 10^{-2})^2 = 17.7\,mWb$

11 Consider a thin 'spoke' of the disc, rotating. The e.m.f. between its ends is:

$E = \Delta\Phi/\Delta t = B\Delta A/\Delta t$

where ΔA is the area it sweeps out in time Δt.
$B = 0.14\,T$ and $\Delta A/\Delta t = \pi r^2 f$

Therefore:

$E = (0.14) \times \pi \times 0.1^2 \times 50 = 0.22\,V$

This is the same for all 'spokes', so is the e.m.f. between the rim and the axle.

12 $E_0 = BAN\omega$, therefore:

$\omega = E_0/BAN = 12/[(0.25) \times (20 \times 10^{-3} \times 35 \times 10^{-3}) \times 300] = 228$

$f = \omega/2\pi = 228/2\pi = 36\,rev\,s^{-1}$

13 For $V = 5\,kV$:

$I = P/V = (20 \times 10^6)/(5 \times 10^3) = 4000\,A$

power loss $P_L = I^2 R = (4 \times 10^3)^2 \times 4 = 64 \times 10^6 = 64\,MW$

For $V = 50\,kV$:

$I = P/V = (20 \times 10^6)/(5 \times 10^4) = 400\,A$

power loss $P_L = I^2 R = (4 \times 10^2)^2 \times 4 = 64 \times 10^4 = 640\,kW$

Note that when the voltage is raised by a factor of 10, the power loss is reduced by a factor of 100.

14 $F = qE = (8.0 \times 10^{-19}) \times (15 \times 10^3) = 1.2 \times 10^{-14}\,N$

$a = F/m = (1.2 \times 10^{-14})/(4.5 \times 10^{-30}) = 2.7 \times 10^{15}\,m\,s^{-2}$

15 $W = qV = \frac{1}{2}mv^2$

$v^2 = 2qV/m = [2 \times (1.6 \times 10^{-19}) \times 1000]/(1.67 \times 10^{-27})$

$v = 4.4 \times 10^5\,m\,s^{-1}$

$v/c = (4.4 \times 10^5)/(3 \times 10^8) = 1.5 \times 10^{-3}$

Final speed is 0.15% of the speed of light.

16 **a** p: proton (baryon), n: neutron (baryon), e^+: positron ((anti)lepton), v_e: electron neutrino (lepton)

b Proton (uud) $\rightarrow$ neutron (udd), so an up quark has changed to a down quark, creating a lepton–antilepton pair.

c $p \rightarrow n + e^+ + v_e$

Charge: $+1 \rightarrow 0 + 1 + 0$ ✓

Baryon number: $+1 \rightarrow +1 + 0 + 0$ ✓

Lepton number: $0 \rightarrow 0 - 1 + 1$ ✓

Index